Responding to the Climate Threat

Gary Yohe · Henry Jacoby · Richard Richels · Benjamin Santer

Responding to the Climate Threat

Essays on Humanity's Greatest Challenge

 Springer

Gary Yohe
Emeritus, Department of Economics
and Environmental Studies
Wesleyan University
Middletown, CT, USA

Richard Richels
Energy Analysis and Climate Change
Electric Power Research Institute (Retired)
Palo Alto, CA, USA

Henry Jacoby
Emeritus, Sloan School of Management
Massachusetts Institute of Technology
Cambridge, MA, USA

Benjamin Santer
Joint Institute for Regional Earth System
Science and Engineering
University of California at Los Angeles
Los Angeles, CA, USA

ISBN 978-3-030-96374-3   ISBN 978-3-030-96372-9   (eBook)
https://doi.org/10.1007/978-3-030-96372-9

This Springer imprint is published by the registered company Springer Nature Switzerland AG
The registered company address is: Gewerbestrasse 11, 6330 Cham, Switzerland

To our children and grandchildren with love: Mari and Courtney plus Katie and Carrie; Mark, Stephanie, Joanne, and Heather plus Sadie, Colman, Preston, Riley, Zoe, Chelsea, Evan, Levi, Sam, Sean, and Cece; Daniel and Caroline plus Margaret, Charlotte, Elizabeth, and Eleanor; Nicholas.

About the Cover

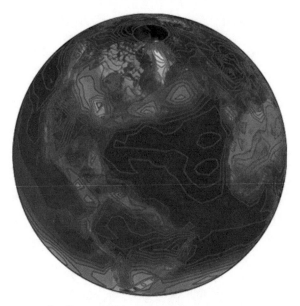

Credit Benjamin Santer (February 2022)

The cover image shows satellite estimates of trends in the temperature of the lowermost layer of Earth's atmosphere (the troposphere). The trends span the 42-year period from January 1979 to December 2020. They are superimposed on a NASA image of the Earth as a "blue marble." Most regions of the troposphere warm during this period. Darker red colors represent larger warming trends. These trends have been plotted so that details of the underlying mountains and ocean basins are still visible. *Climate scientists have shown that this observed global pattern of warming of the lower atmosphere cannot be explained by natural causes alone—and so it represents a clear "fingerprint" of human effects on climate.* Mitigating this warming will be "humanity's greatest challenge" in the 21st century. The satellite data used in generating the image were provided by the Center for Satellite Applications and Research in Camp Springs, Maryland, USA.

Preface

Our joint objective in writing these essays and preparing this annotated collection has been to help inform the public and policy-makers about the threat of climate change, writing for the general reader but not watering down science and policy complexities. To this task we have tried to bring to bear our experience in teaching and research on climate change, and what we learned from participation in scientific and policy assessments over the years.

We came to this activity by a circuitous route. Our collaboration began in the summer of 2018 when Rich called Gary about a possible collaboration. Rich wanted to write a humorous narrative attacking those sowing false information about climate change. The idea was to construct a fable based on The Emperor's New Clothes, with then-President Trump as the emperor. Jake first entered by commenting on an early fable draft, but soon joined the team. It was in this period that we began regular weekly phone meetings—a highly productive practice we have continued since, now in the form of much anticipated Zoom sessions.

Our attempt at sarcasm got some exposure, but we soon concluded that perhaps what talents we have were better suited to non-fiction. We agreed we would take on any climate-related topic that was relevant to the public dialogue, on the condition that we would stay within the "sandbox" of our own disciplinary knowledge. That is, we would take care not to overstep the boundaries of our specific education and experience, risking mistakes that would undermine the credibility of the effort.

In early 2020 the team expanded again. We were limited by our disciplinary training—in the social sciences, mainly in economics—and we turned for advice to Ben Santer, an atmospheric scientist and long-time IPCC colleague. Fortunately, it turned out that Ben was interested in our project and could provide a much-needed expansion of expertise and focus into the physical sciences. He joined the group and the "gang of four" was complete.

We never lacked for topics in our weekly phone or Zoom meetings. Also, each of us welcomed their therapeutic value. Those were stressful times—individuals in the Trump Administration and elsewhere attacked our life's work, incorrectly dismissing climate change as a hoax. The experience led us to wonder: Why hadn't we begun working together on this crucial communication task years ago?

Our essays collected here are those published beginning in 2019, continuing through to the middle of 2022. They convey the state of climate science, economics, and policy as we knew it at the time. The volume is organized into seven parts. We start with Part I, providing motivation and an overview, and close with a discussion in Part VII of the task ahead for climate science communicators. Parts II through VI contain the essays, which are arranged approximately but not exclusively in chronological order. For each essay, we provide a Prologue that explains what led us to take up the topic and an Afterword that considers what happened after the essay appeared.

Our focus throughout has been primarily on the U.S. Beginning in 2019 the U.S. Presidential primaries loomed large in public consciousness, and our essays from this period are collected in Part II. In early 2020 the COVID pandemic struck. The parallels with climate change were evident and led to the essays in Part III. In the run-up to the 2020 election we produced a series of essays rebutting climate denial and providing positive evidence of strong and damaging human influence (Part IV). In Part V, we reflected on issues faced by the new Biden-Harris Administration. Finally, the essays in Part VI we considered ongoing challenges in the fight against climate change.

We engaged in long debates about how best to organize the essays in one volume. It turned out not to matter, though, because we ultimately decided that we should have no expectation that you, our reader, would necessarily approach these essays from front to back—or in any other particular order for that matter. We do hope, however, that the essays will provide you with the information and the sense of urgency to be an effective participant in the struggle against global climate change.

We believe that it is now more urgent than ever to translate complex science and policy issues for the lay public. This book is a beginning for us four, not an end. The opportunity to educate about the reality and seriousness of climate change, and help eliminate our common threat, is never over.

Middletown, CT, USA   Gary Yohe
Cambridge, MA, USA   Henry Jacoby
Summit, NJ, USA   Richard Richels
Klamath Falls, OR, USA   Benjamin Santer
November 2022

Acknowledgements

We four acknowledge decades of contributions arising from collaborations with many colleagues in our home institutions, the Intergovernmental Panel on Climate Change (IPCC), the U.S. National Climate Assessment (NCA), the National Academies of Science, Engineering, and Medicine (NAS), the National Research Council (NRC), New York (City) Panel on Climate Change (NPCC), and many other venues. Any remaining errors are ours.

Except for nominal compensation from the Guardian for Essay 23, donated to the National Center for Science Education, no external funding was received for any of the essays. Isabelle Harper, a rising senior in the Greenwich CT High School in 2020, was engaged to help edit our text. She worked to clarify our jargon, alerting us to references that made sense to us four senior citizens but were perhaps incomprehensible to Millennials.

Many friends helped us in this effort, but we are particularly indebted to Ed Maibach of the George Mason University Center on Climate Communications for his wisdom about how to write for a general audience, and to Bud Ward of Yale Climate Connections for his consistent guidance on when particular issues would be salient (and for his keen editorial eye when we were just getting the hang of writing these pieces). Michael Oppenheimer spent a long hour one winter Tuesday bringing us up to speed about tipping points and sea level rise. Eric Rignot from the University of California at Irvine took time out during a trip to Rome to follow up with some detailed information about the international modeling and data collecting research efforts in Antarctica, with a specific focus on rapid changes in the Thwaites glacier.

We also wish to thank the readers of the originally published essays who often provided us with valuable feedback—and sometimes let us know when we had touched a nerve one way or the other. We all need more of that.

Contents

Contents   xvii

Abbreviations

ACC	America's Climate Choices
CCL	Citizens' Climate Lobby
CDC	Center for Disease Control (and Prevention)
CFC	Chlorofluorocarbon
CNN	Cable News Network
CO_2	Carbon dioxide
COP	Conference of the Parties (of the UNFCCC)
COVID or COVID-19	The novel coronavirus
EPA	Environmental Protection Agency
EPRI	Electric Power Research Institute
EU	European Union
FDR	Franklin Delano Roosevelt
GHG	Greenhouse gas(es)
GND	Green New Deal
IPCC	Intergovernmental Panel on Climate Change
MIT	Massachusetts Institute of Technology
NAS	National Academy of Science
NASEM	National Academies of Science, Engineering and Medicine
NCA	National Climate Assessment
NIH	National Institutes of Health
NOAA	National Oceanographic and Atmospheric Administration
NPCC	New York (City) Panel on Climate Change
NRC	National Research Council (U.S.)

OSRD	Office of Scientific Research and Development
OSTP	Office of Science Technology and Policy
RFC	Reasons for Concern
SCC	Social cost of carbon (or carbon dioxide)
SSRN	Social Science Research Network
STEM	Science, Technology, Engineering, and Mathematics
U.S.	United States
UNFCCC	United Nations Framework Convention on Climate Change
USGS	United States Geological Survey
WCRP	World Climate Research Program
WGI or WGII or WGIII	Working Group I, II, or III in an IPCC Assessment
WHO	World Health Organization

Part I Motivation and Overview

Anyone interested enough to crack open this book is no doubt aware of the seriousness of the climate threat, and that action by the U.S. is essential to any global solution. There is no international police agency to enforce the needed reduction in greenhouse gas emissions. The global agreement to meet the climate threat, embodied in the Paris Agreement, involves a system of voluntary "contributions" by individual nations. In such a regime, foot dragging by the United States—the richest country, and the largest source of greenhouse gas emissions to date—dampens the incentive of other nations to take action. It follows that climate action by the U.S. is a "two-fer". It would both reduce the U.S. contribution to global emissions and (equally important in our view) lower global emissions by encouraging similar efforts elsewhere.

In these essays, and by gathering them in this book, we hope to contribute to this crucial effort. Our instrument in this fight against climate change is the written word. Our efforts fall in a space in between the professional literature on climate issues and articles by staff writers in newspapers, magazines, and electronic media. Indeed, we hope to not only provide clarity for the lay reader but also, occasionally, to help correct misunderstanding or misinformation in the daily news feed.

To provide an overview of some of the problems that underlie the challenge of public communication and its effect on understanding of the climate threat, we begin here with a summary of some of the sources of hesitancy to take climate action. These sources become targets throughout the series of essays. We then explain the five-part organization of the essays, and we

© The Author(s), under exclusive license to Springer Nature
Switzerland AG 2023
G. Yohe et al., *Responding to the Climate Threat*,
https://doi.org/10.1007/978-3-030-96372-9_0

provide a very brief summary of what each essay is about—supplemented by a comment on ways a reader might productively approach the volume.

As an additional aid in reading the essays, we offer a description of the mindset we bring to the climate change threat and policy choices. We present our view that climate change is the ultimate societal challenge—the stakes are enormous, maybe inestimable in the long term. Effectively dealing with this challenge requires the balancing of risks: that of an inadequate response to the threats posed by an overheated planet, leading to unnecessary pain and suffering, versus the risk of acting too quickly or with wasteful policies, thereby squandering valuable resources that could be better applied to other societal and environmental problems. Our focus is on mitigation and adaptation decisions to be made over the next decade. These decisions must be reached in the face of longer, century-timescale scientific uncertainties, recognizing that there will be opportunities for learning and mid-course corrections.

Impediments to Public Support

Why is it taking the American public so long to understand the climate threat and support a more aggressive response? Are they unaware of the overwhelming body of scientific evidence pointing toward the reality and seriousness of human-caused climate change? Have too many U.S. citizens accepted the false narratives being peddled by well-funded climate deniers? Is the public just too absorbed with what they felt were more immediate concerns? The answers to these questions are important: policy-makers cannot get too far ahead of public understanding of the climate threat, or of public perception of the priority for needed government action.

There are, of course, many reasons behind the public's slow response, or even opposition to any action at all. But it seems to us that four phenomena tend to come up again and again, and so are targets for more effective public communication.

Lack of technical information targeted to a public audience. Scientists have taken the climate threat seriously for decades, publishing widely about it. Some have addressed individual pieces of the climate puzzle: for example, the various components of the carbon cycle, the magnitude of feedbacks in the climate system, or the characteristics of specific energy technologies. Other studies have been broader, focusing on the chain of events that link human activities to warming, how damages may grow and be distributed geographically, and the necessary conditions for a successful agreement to reduce global greenhouse gas emissions.

Yet despite the widespread availability of this information, much of the science of climate change and the complexities of policy formulation remain impenetrable for non-technical audiences. Many members of such audiences often lack the time or inclination to dig beneath the surface—they want to get to the bottom line quickly. How will climate policy affect their families, their livelihoods? Physical and social scientists have attempted to answer these questions, but too little of their writing is readily accessible to readers unfamiliar with the disciplinary jargon. Help is needed to connect the dots.

Difficulty incorporating scientific uncertainty. People deal with uncertainty all the time. We all make decisions about what to do in response to some possible future event. Our responses are influenced by the likelihood of the event, and by how bad or good the consequences might be if it actually happened. For example, in fear of a possible house thief, many people lock their doors. Others install an alarm system, live behind a protective gate, buy a guard dog, or even hire a permanent security detail. All of these responses depend on our individual perception of the level of risk and the resources at our disposal.

A natural way of thinking about such choices is from a risk comparison perspective. For example, in the face of a negative event, we may balance the cost of different levels of avoidance against the lowered likelihood of the potential future loss. This way of thinking, however, does not seem to easily carry over to the threat of global climate change and its associated economic and environmental damage. Uncertainty in projected climate damage becomes a justification for doing nothing or waiting for better information.

In part, this "paralysis by uncertainty" is due to distrust resulting from inadequate understanding of how estimates of expected climate change and its effects are produced. To the lay person, it can appear that results appear mysteriously from some black box. In fact, projections of climate change and its impacts are based on knowledge of the heat-trapping properties of greenhouse gases, evidence from global observation of hundreds of different properties of the climate system, and analysis in virtual laboratories for studying Earth's climate. In the latter case, these virtual laboratories consist of computer models of the land, atmosphere, ocean, and cryosphere.

Rather than emerging from a black box, climate science is based on a combination of basic theory, rigorous observations, and computer modeling. All this research is carried out by hundreds of independent groups around the world. The breadth and depth of these scientific efforts—and how they have evolved over time—is not common knowledge. A greater effort is needed to fill the gap between the scientific literature and day-to-day news coverage. The goal is to enhance awareness and understanding of the richness and internal consistency of the evidence for human influences on global climate.

Distrust of the scientific enterprise itself. Slow buy-in to the findings of climate research and analysis may also reflect a more subtle and insidious influence: a deep-seated distrust of science and the institutions that fund it. Climate change is not alone in this difficulty, as is clear from the all too common rejection of medical advice during the COVID pandemic. Lessening this distrust will depend on a deeper public understanding of the scientific community and how it works. The notion that science is performed by some obscure government office needs to be dispelled. In climate science, understanding arises from a large, diverse, and international collection of federal and state laboratories, university research groups, non-governmental organizations and private corporations. Research conducted by these groups is assessed in an open and transparent way, in full view of many different stakeholders.

Climate Denial. We make a distinction between skepticism and denial. The former is at the heart of the scientific method. It drives conjecture, hypothesis testing, and the creation of knowledge. Conversely, the term "denial" is used when the rigorous scientific analysis required for enhancing the knowledge base is omitted or ignored.

We encounter two troubling forms of climate denial. First, there are those who reject out of hand any findings that are inconsistent with their preformed conclusions. They are already convinced that climate change is a "hoax" and only listen to those with similar world views. This group is unlikely to change their opinions in response to better communication of the science or reliable information about the advantages of a particular policy response. We don't see this group as our primary audience—although we always hope to trouble them with hard scientific evidence. Rather, we write for those who know that global warming is a problem, who seek a deeper understanding of causes, effects, and productive responses, and who might welcome a more informed basis for responding to the arguments of their denier friends.

Then there is another form of denial that is of growing concern. It does not reject the science of climate change. Instead, it lapses into despair, concluding that society is unable to limit the increasing emissions of greenhouse gases—the primary cause of planetary warming. Holders of this view believe that it is not worth the effort to devote resources to emissions control. It is better for the economy, and psychological health, to "go ostrich" on emission control and focus instead on adapting to change that we can't stop. In communication with this group, there is need for improved dialogue about the natural and social science of potential emissions control measures. It is essential to promote a better understanding of what can actually be achieved, and to be clear on why "sooner is better" in terms of emissions reduction. Our capacity to adapt may eventually be overwhelmed by climate change. Adaptation and mitigation must go hand in hand.

The essays in this volume are our effort to help remove these impediments to action on climate change. Our approach has been to distill messages from across climate-related natural and social science and to make hard-won insights more accessible to politically active citizens. In the following, our focus is primarily on the U.S., since we are most familiar with the U.S. policy landscape which many of these essays deal with.

The Essays—A Summary

Just as in comedy, meeting the objectives described above is all about timing. Opinion page editors have short attention spans. Their interest in new material is limited to the topic of the day. The willingness of readers to pick up an article depends on which items out of a vast stew of public issues are salient at any particular moment. Catalysts of interest can be climate-related events, but opportunities to inform about climate change can arise when the public is focused on an issue like the COVID pandemic, which also involves science and decision-making under uncertainty.

The issue of timing is evident in the ordering of individual essays. The essays are organized here in Parts II through VI. They are in chronological order, with one exception. Part II begins with essays written early in the 2019-20 political season, Parts IV and V contain articles written in the weeks immediately preceding the 2020 election and in the years thereafter. Essays in Part VI focus on continuing challenges. The exception to this rough chronology is in Part III, which contains essays written in response to the COVID pandemic.

Each essay has a Prologue explaining the context for the essay and why we decided to write it. Essays are also accompanied by an Afterword discussing what happened after the article appeared.

Part II. Climate Change and the Political Environment. The first block of eight essays appeared in 2019 and the first months of 2020. Neglected for years by the Trump Administration, the climate issue began to take on greater salience with the approach of the 2020 Presidential election. In

February 2019, for example, Senator Markey and Representative Ocasio-Cortez attracted media attention and controversy with the introduction of a congressional resolution on the Green New Deal. The celebrity of Greta Thunberg was growing as she gave voice to the dangers of the climate threat and the unfairness of climate inaction. The renewed focus of attention on climate was an opportunity to add to public information on the environmental threat and state of the public response to it. Early essays addressed the Green New Deal and climate issues in political debate (Essays 1 and 2) and a common misunderstanding of U.S. national interest in the issue (Essay 3).

Two subsequent essays deplored the low priority accorded to climate in the U.S. (Essay 4) and the lack of urgency in international climate negotiations (Essay 5). Hurricane Dorian provided an example of the implications of inaction (Essay 6). Highlighting the importance for climate policy of U.S. electoral choices, two final essays in this part call attention to a Trump Administration policy change hampering climate action (Essay 7), and to President Trump's continued denial of the threat (Essay 8).

Part III. *Climate change and the COVID-19 virus*. In the early months of 2020, the pandemic hit, dominating public attention. Taking advantage of the heightened focus on a serious threat to the health and stability of global society, we composed a second block of essays. We attempted to teach about climate by highlighting the similarities, differences, and direct connections between the climate and COVID threats. They may differ in time frame—COVID ramped up over weeks and months, while human-caused climate change has evolved into a critical threat over decades—but for both early action is crucially important for an effective response.

In exploring the direct connections between the two threats, one essay considers how the response to COVID may or may not help spur action on climate (Essay 9) while another explains how climate change increases both the likelihood and damage of pandemic disease (Essay 10). In addition, controversy over published analyses of COVID policy provided an opportunity to explain how science works—it is dynamic, not static, and routinely explores the implications of different policy choices (Essay 11). Finally, the word "existential" is applied to both the COVID and climate threats, often with some confusion, and we took this opportunity to provide clarification (Essay 12).

Part IV. *The Yale Project for the Campaign Season*. In the summer 2020, as the presidential campaign was heating up, we began a conversation with Yale Climate Connections (YCC). It is a non-partisan service dedicated to helping citizens and institutions understand the climate threat. We shared with YCC leadership our concern about President Trump's persistent climate denial, and we fully expected that global climate change would be on the 2020 ballot. It was important, therefore, to counter the disinformation that likely would be broadcast during the 2020 campaign. Our response was this block of essays, which YCC kindly agreed to publish.

We began by suggesting questions that might be used by moderators in the presidential debates (Essay 13). Our next product was a set of six articles countering some of the principal arguments of climate deniers, published one per week in the run-up to election day. We started with the argument that climate change is not happening (debunked in Essay 14) and proceed to address the false claim that even if climate change is happening, humans are not responsible (countered in Essay 15). We then moved to the assertion that the rate of climate change is not concerning and there is no basis for expecting climate change will get worse (debunked in Essays 16 and 17). In the penultimate Yale essay, we took on the fallback denier talking point that even if humans are causing climate change and it's bad, the proposed control policies would be too expensive (Essay 18). Finally, we responded to the incorrect argument that the U.S. is only responsible for a small part of current global emissions (around 15%), so it is not in our national interest to get out ahead of other nations in taking action (debunked in Essay 19).

Part V. Early Months of the Biden Administration. With the 2020 U.S. election results in and a new administration assuming the reins of power, the mood and the communication challenge shifted. President Biden accepted climate science, understood the threat of human-caused climate change, and was committed to action. Getting effective climate policies in place would not be easy, however. Deniers were still muddying the waters and needed to be countered. However, we saw the need at this time to focus first on federal climate actions, to explain their importance, and to counter arguments opposing effective action. We also kept an eye on the continuing damage caused by extreme weather events.

Four of these essays are arguments for potential components of climate policy. One concerns a particular way to think about dealing with the climate threat, which is as a problem of risk management. We argue that this way of framing the challenge should pervade the National Climate Assessment, a congressionally mandated study of the climate problem that takes place every four years (Essay 20). The next two essays concern congressional and executive actions. One essay urges the Biden Administration to watch the clock, and not run past the deadline by which some end-of-term executive actions by the Trump Administration could be reversed (Essay 21). The other observes the broad public support for climate-related executive actions by the new Biden Administration (Essay 22). The last of these four explains the advantages of a carbon tax, a measure not on President Biden's agenda (Essay 23).

We also call attention to the Biden Administration's renewed respect for science, comparing the challenge it faces today with a similarly challenging time in the aftermath of World War II (Essay 24). In contrast, Essay 25 provides a critical review of a book that takes a very different view of the credibility of climate science.

Meanwhile, the climate took no notice of these efforts. Climate-related damages continued to mount. We took another stab at explaining how these extreme events are related to global warming (Essay 26) and then summarized the ever-growing list of events that had "never been seen before" (Essay 27).

Part VI. The Continuing Challenge. This part of the volume begins with good news: a description of progress in one of the great scientific challenges: understanding how Earth's climate system works. Essay 28 describes how these advances have led to ever more confident statements (e.g., by the IPCC) of the human cause of warming. Subsequent articles returned to both old challenges and to the new ones that are continually arising. One new emerging challenge is the fragmentation of the global system, erecting new barriers to cooperation on climate (Essay 29). Then, reflecting on the Ukrainian war, Essay 30 points out how the global response to the resulting energy crunch has implications not only for the climate but also for the future of democratic institutions. Another long-standing "forever" challenge is to devise a climate response that doesn't exacerbate existing domestic and international inequities (Essay 31). Still another difficulty arises in maintaining the proper balance between ambition and reality in setting goals and expectations (Essay 32). Essay 33 closes this set by arguing that temporary or easily reversed measures, undertaken when political support is weak, will be insufficient to meet the climate change challenge.

Reading This Book

We don't expect that you will read this collection of essays in the order presented here. We hope that this introduction to the essays guides you to where you would like to start. Every part is self-contained, and each essay stands alone in content and context. It is, therefore, a collection of connected short reads. If our experience authoring each essay and its surrounding text is any indication, however, it is not good bedtime reading.

How We Think About the Problem

You may be wondering how we created the present climate problem. The process is all too easy to see. Start with the industrial revolution when much of the world first became hooked on fossil fuels. Add the internal combustion engine and fossil-fired power plants and long lags between cause and effect. Add population growth and urbanization. And add special interest groups and a political system which caters to them. Then add mitigation actions that involve large upfront costs and delayed payoffs. Add a developing world seeking economic parity with its wealthy neighbors. And add, of course, eight billion citizens with different perspectives on what matters to them and to future generations. Taken together, all this complexity means that coping with human-caused climate change is a fiendish problem of global dimensions. And there is no simple fix.

So how do we begin to clean up this mess? The first step is to get our arms around the problem. Here and throughout the essays, we adopt a risk-based approach that we have found useful in organizing our own thinking. It is nothing new. Risk is likelihood times consequence, adjusted for one's aversion to a particular scenario. It has been a fundamental part of the core curriculum of environmental science students for generations.

The risk-based approach requires weighing climate risks with the risks and costs of potential response measures in the search for sensible steps to be taken now. We must consider the errors in doing too little to abate and adapt. We must look equally carefully at the potential mistakes in trying for too much, too fast, and squandering valuable resources. This type of approach is known by many names. It varies from discipline to discipline. Some refer to it as risk-benefit-cost analysis[1]. We call it risk management.

But risk to whom? Being global in scope, there is no single decision-maker who determines the level and distribution of mitigation, adaptation, and suffering. There is only a collection of countries—a hundred and ninety plus—loosely knitted together by a common objective. Much of what we have written is from the US perspective. It is, as we will argue, the country whose participation and leadership is essential to meeting the global climate goals—and the country bearing the most responsibility for historical emissions.

In what follows, we set the stage for thinking about the climate problem. We first describe its bewildering nature, then talk about the potential harm

---

[1] Fischhoff, B. (2015), 'The realities of risk-cost-benefit analysis', Science 350(6260): 527.

to the inhabitants of the planet, and follow with a discussion of the need for understanding and highlighting the tradeoffs in different courses of action.

The ultimate tragedy of the global commons. Garrett Hardin used the metaphor of a herd of cattle overgrazing a public commons in his classic article published in 1968.[2] He observed that when it comes to common resources that are in limited supply, individuals will often neglect societal well-being in the pursuit of personal gain, to the detriment of all. The "commons problem" is likely to arise not only with shared grazing grounds, but also in a host of other settings involving the communal use of resources, including forests, rivers, oceans, and the atmosphere. Climate change is sometimes referred to as "the ultimate tragedy of the global commons" and the "ultimate challenge for economics."[3]

Any solution will necessarily be inherently political in nature. Attempts to halt global warming have been long on good intentions but short on real progress. The 1997 Kyoto Protocol is a case in point. The focus of Kyoto was on mandatory commitments for industrialized countries, and that approach failed. The 2015 Paris Climate Agreement marked a major change in emphasis. Its objective is voluntary emissions reductions by all. Each nation has been asked to specify its contribution to holding average global temperature "well below" two degrees Celsius (2°C) relative to preindustrial levels, and to try to limit the change to 1.5°C. Although concerns remain about the practicality of a purely voluntary approach, this has been the basis of virtually all past environmental agreements. It offers perhaps the best chance for widespread agreement on how to achieve progress. But will it be sufficient to achieve net-zero GHG emissions?

The changing character of the climate problem. Veterans of the climate wars must yearn for the days when the issue was whether a warming planet would be beneficial for its inhabitants—as Svante Arrhenius, a Swedish Nobel Laureate, postulated more than 100 years ago. His thesis was that warming would increase food production and help feed a growing population. The science of climate change has advanced substantially since then.

Initially, reasons for concern focused on such issues as species extinctions, coral bleaching and mortality, the survival of tropical rainforests, sea level rise, ocean acidification, and the increased vulnerability of indigenous communities in the Arctic and on small islands. But more recently, attention has expanded to include extreme climate events. The dramatic upsurge in devastating wildfires, extreme Atlantic hurricanes, floods, record-shattering

---

[2] Garrett **Hardin**. *Source* Science, New Series, Vol. 162, No. 3859 (Dec. 13, 1968), pp. 1243–1248.
[3] Nordhaus, W.D., Nobel Lecture, 2018, https://www.nobelprize.org/prizes/economic-sciences/2018/nordhaus/lecture/.

heatwaves, and severe droughts and famines has exacted a substantial toll on both ecosystems and human systems.

Not only does the likelihood of such events rise with increasing temperature—the severity of damages does as well. Studies conducted by the National Oceanic and Atmospheric Administration (NOAA) have carefully monitored the statistics of extreme events over the past four decades.[4] The record of changing extremes is both irrefutable and concerning. Even more worrisome, the scientific evidence strongly suggests that in the absence of decisive mitigation activities, what we are now seeing portends an all too plausible and devastating future. We have seen the canary in the coal mine—a troubling omen of even worse to come.

Present concern is not confined to extreme events that make the evening news. A growing number of scientists fear large-scale discontinuities in the climate system that could result in a domino effect as multiple "tipping points" are crossed. Many worry that the rise in global temperatures will accelerate as ice sheets disappear, forests weaken and decline, permafrost melts, and sea levels rise. Oceanographers are concerned that the circulation in the North Atlantic, an important part of the planet's heat distribution system, will be drastically altered. Although the existence of dangerous tipping points in the climate system is known from the study of paleoclimate data, we do not know how close we are to them.

The changing character of the climate debate. Damages are only part of the picture. There is a natural tension between those who focus on the impacts of global warming and those whose attention is drawn to the costs of abatement. At its inception, the drafters of the U.N. Framework Convention on Climate Change (UNFCCC) worried that antagonists would successfully manipulate analyses of costs and benefits of action to make the case against halting global warming. They feared that future benefits would pale in comparison to upfront abatement costs due to the vagaries of discounting. To discourage such mischief by numbers, the ultimate objective of the UNFCCC was clearly articulated: the prevention of "dangerous anthropogenic interference with the climate system." This is the motivation for the 1.5°C temperature limit and the associated greenhouse gas concentration window suggested by the IPCC. Costs come into play only in that "measures to deal with climate change should be cost-effective so as to ensure global benefits at the lowest possible cost."

Not surprisingly, such instructions are frequently disregarded by those choosing to do so. The anti-abatement proponents unabashedly claim that

---

[4] https://www.ncdc.noaa.gov/billions/.

proposals aiming to radically cut back greenhouse gas emissions would be exorbitantly expensive, harming the U.S. economy, costing millions of jobs, and leading to severe recession. The reduction in harm to health and well-being is ignored or assumed to be non-existent. In any event, it is left out of the calculus.

In response, those arguing for a rapid phasing-out of fossil fuels point out that yes, abatement costs may be substantial, but pale in comparison to the far greater damage costs. Indeed, proponents of emissions reduction say that the biggest threat to economic prosperity will come from continuing to ignore the threat of climate change.

As the implications of not controlling GHG emissions have come into sharper focus, we need not fear that weighing costs and benefits will lead to further procrastination. While this may have been the case three decades ago, it is not so today. We have misjudged the magnitude and timing of damages and the lead-time necessary for action. The dynamics of the human and physical systems suggest that the time for procrastination is over. Pain and suffering from further procrastination will not only fall on the unborn—it will also affect the current inhabitants of the planet. Those who believe that their sacrifices are being made only for the benefit of far distant generations need to think again. The unprecedented risks are there for all to see. Level-headed assessment of these risks is what is needed for informed public policy. Rather than ignore the tradeoffs between damages and abatement costs, these tradeoffs need to be made clear to all.

This is the way we see the problem.

Part II

The Gathering Storms

In 2019, the U.S. political season was well under way in the run-up to the 2020 election. We expected that climate change would be an important issue. Opinion polls were showing a rising percentage of respondents who accepted that climate change was real and supported government action in response. It was an opportunity for short articles on various topical issues related to climate.

We began with two articles motivated by then-current political events—one expressing our concern about a problematic climate policy proposal (Essay 1), and another offering suggestions of how climate issues might be framed in the electoral-season debates (Essay 2). But we soon shifted to focus on deeper problems impeding U.S. and international efforts to respond to the threat of climate change. One essay attacked the common use of the phrase "America First!" for its lack of appreciation that it is in the U.S. national interest to support the climate change programs of less fortunate nations (Essay 3). Two other essays addressed our seeming inability to make progress in fighting this global problem—the low priority that is given to U.S. action on climate change, even among people who accept it is a problem (Essay 4), and the lack of a sense of urgency at a Madrid meeting of the Climate Convention—as if the planet could wait for us to figure out what to do (Essay 5). In the summer of 2019, Hurricane Dorian provided a dramatic example for an essay in which we highlighted the damage to come if nations fail to mount an urgent, global attack on the problem (Essay 6).

Two other events—the Administration's slashing of a federal estimate of the seriousness of the climate change threat (Essay 7) and President Trump's dismissive response (Essay 8) to devastating California wildfires—provided motivation for essays highlighting the enormous stakes involved in the upcoming 2020 national election.

Essay 1. Don't Let the Green New Deal Hijack the Climate's Future

Prologue

There is a continuing need to raise climate change to greater prominence in U.S. political discussion, and in 2019, a congressional resolution was introduced by leaders in the House and Senate to do just that. Termed the Green New Deal, it was a grand plan for tackling climate change. The Green New Deal was linked to a series of other social objectives, as suggested by the reference to Roosevelt's New Deal of the 1930s. Fairness will be a requisite of any successful climate strategy, but the Green New Deal seemed a leap too far. We wanted to warn that any associated push for climate action would likely be lost in opposition to the proposal's larger social agenda.

This essay first appeared in *The Hartford Courant*, March 1, 2019.

© The Author(s), under exclusive license to Springer Nature Switzerland AG 2023
G. Yohe et al., *Responding to the Climate Threat*,
https://doi.org/10.1007/978-3-030-96372-9_1

Don't let the Green New Deal hijack the climate's future

Gary Yohe, Richard Richels, and Henry Jacoby

March 1, 2019

The Green New Deal calls for the country to meet a number of ambitious environmental targets while solving a host of other ills facing our country. Its near-term environmental goals, authored primarily by U.S. Rep. Alexandria Ocasio-Cortez, D-N.Y., and U.S. Sen. Ed Markey, D-Mass., include "meeting 100 percent of the power demand in the United States through clean, renewable and zero-emissions energy sources," "building or upgrading energy-efficient, distributed and 'smart' power grids." "working to ensure affordable access to electricity" and "overhauling transportation systems in the United States to eliminate pollution and greenhouse gas emissions … as much as technologically feasible."

These goals are aspirational, but they are all issues that are worthy of serious and urgent consideration.

It is likely that the authors of the resolution and its co-sponsors thought that yoking climate change to their larger agenda would start conversations. But this version of the Green New Deal will impede continued progress on the climate front for years to come.

Does one have to support and defend all of the goals of the Green New Deal to push forward on climate change? If the answer is yes, then past experience teaches that political manipulation will *doom* climate change to artificial partisan firefights for years to come.

Around the turn of 2019, climate change had become a rising star in the national political galaxy. The connection of recent catastrophic wildfires to increased temperatures had become clearer. So had the link between warming temperatures and changing atmospheric currents to more severe hurricanes. The link to sea level rise had been solidified. Assessments released late last year brought attention to climate risks in 2030 — not 2100. Candidates for Congress had run on platforms that highlighted climate change as an existential issue — and won. Media coverage had begun to report that climate change would likely to play a significant role in the 2020 presidential campaign.

But now, we see calls for climate policy lumped together with calls for universal education, universal high-quality health care, guaranteed affordable and adequate housing, economic security and access to healthy and affordable food. Opponents can now label the entire collection of worthy social objectives as modern "socialism."

As a result, we see honest climate concerns fading from the national conversation. Instead, the "ice cream" concerns of Sen. John Barrasso , R-Wyo., and the president's fear of losing "airplane rights" are the talking points. The Republicans' objective here is to create hyperbole images of how the Green New Deal would radically change the United States.

A more responsible and focused approach would have involved posing questions like: What will the climate and climate politics look like in 2030? What additional damage will have been attributed to anthropogenic sources? Will the United States have been dormant, or will we have re-engaged in responding effectively to climate's immediate and existential threat?

We think that a dormancy future of no action could be more likely. We now envision a much heavier lift in achieving climate action at the federal level over the next decade.

A more optimistic view of 2030 would not rely on revolutionary change in the United States or around the world. The *beginning* of a revolutionary transition in energy supply and use and in land management techniques *would, however,* be essential; and that will require federal support.

This more positive vision offers a less hostile future in 2030. But lapsing into the pessimistic future stoked by the "socialism" label, the U.S. federal role in this decades-long transition would start many years later— a very costly outcome.

For the sake of our climate, it will be very important that our political leaders guide all of the issues raised in the Green New Deal to their proper political domains — paying careful attention to positive interactions among them, but *not* linking them in one grand national campaign.

For the sake of our children and generations to follow, progress on climate change must not depend on an up or down vote on the entire Green New Deal agenda. Conversely, carefully crafted action on climate change can help underwrite progress on its broader social objectives.

Afterword

It was not long before the label of "socialism" was attached to the resolution, with arguments that it was "a big-government takeover of the economy masked as an environmental policy." The plan was never put forward in legislative language. Within a year, the Green New Deal's momentum on social issues was swallowed up in the pandemic. Attention turned to the tattered social safety net that COVID-19 revealed, and to legislative efforts to moderate COVID's damage.

The concept of a Green New Deal has not gone away, however. Nor have the social ills that were a large part of its motivation. The name sticks, and components of the original proposal continue to be featured in the progressive agenda. Indeed, the task that the essay addressed—to keep a focus on the climate threat while recognizing its intersection with other national problems—will be a continuing challenge.

Essay 2. Advice on Climate Policy for the 2020 Presidential Candidates

Prologue

In the summer of 2019, there were over a dozen candidates running for the Democratic nomination for the U.S. presidential election. Debates between these candidates were under way. We hoped that they would be asked about their positions on climate change in these events. We hoped they would address the issue in their stump speeches. So far as we knew, all the Democratic candidates were aware of the climate threat and supportive of government action. It was not clear how they would handle the topic and its controversies in public debate. From our own missteps, we had some idea of where the pitfalls lie in presentations to general (and perhaps hostile) audiences. And so, despite our inexperience with political campaigns, we offered some advice on more effective communication approaches, and points to avoid.

This essay first appeared as https://thehill.com/opinion/energy-environment/453755-advice-on-climate-policy-for-the-2020-presidential-candidates.

Advice on climate policy for the 2020 presidential candidates

Richard Richels, Henry Jacoby, and Gary Yohe

July 18, 2019

The Paris Agreement,[1] with its goal of halting global warming short of 2 degree Celsius, conjures up an image of a temperature threshold which we dare not exceed, akin to Thelma and Louise knowingly and recklessly driving over a cliff to their ruin.[2]

For that hapless couple, there may have been a sudden abyss that they chose to breach. But in the case of global warming, a better metaphor might be that of the can that gets kicked down a road that becomes increasingly treacherous with every mile travelled. Due to our past reluctance to apply the brakes, the can is now farther down the road, and it is going faster than we realize.

The science is clear. When we include the pent-up momentum of the climate system, we have already committed to warming of at least 1.5 degrees C.[3] Moreover, with the additional heating that will occur as we reduce emissions to zero, a 2-degree limit is also in doubt.

Have no illusions. The case for coordinated action both nationally and internationally is compelling, even if the articulated temperature targets turn out to be only aspirational. The intensity of the wildfires in the West,[4] the unrelenting flooding[5] in the midsection of the country and the fury of hurricanes striking[6] our coastlines are but a preview of what's to come—just like the recognition of increases in the frequency of heatwaves and the severity of droughts.

It doesn't take a genius to see the absurdity of inaction in the face of these risks. Continuing to kick the can down the road will place an intolerable burden on future generations.

But all the news is not bad. Recent polls suggest that we may be entering a new era of public concern[7] over climate change. The number of Americans witnessing[8] the growing destruction has risen. Many see it out of their kitchen windows; all observe it on the evening news.

---

[1] https://unfccc.int/process-and-meetings/the-paris-agreement/the-paris-agreement.

[2] https://www.imdb.com/title/tt0103074/.

[3] https://www.ipcc.ch/sr15/chapter/summary-for-policy-makers/.

[4] https://www.washingtonpost.com/business/2018/08/14/wildfires-have-gotten-bigger-recent-years-trend-is-likely-continue/.

[5] https://www.nytimes.com/2019/05/24/us/midwest-river-flooding.html.

[6] https://theconversation.com/hurricanes-to-deliver-a-bigger-punch-to-coasts-113246.

[7] https://www.vox.com/2019/1/28/18197262/climate-change-poll-public-opinion-carbon-tax.

[8] https://climatecommunication.yale.edu/publications/a-growing-majority-of-americans-think-global-warming-is-happening-and-are-worried/.

Moreover, many state and local governments,[9] corporations[10] and individuals are joining the worldwide effort to halt carbon pollution. Although no substitute for U.S. leadership at the national level, these efforts may at least keep our metaphorical can in sight.

It is at the national level where action is most conspicuously absent. Bipartisan congressional support still will be essential even with a more sympathetic president. The question is how to obtain it. Perhaps, once the public becomes convinced that the unprecedented damages[11] being observed are a preview of worse to come, they will demand action. And if the politicians believe they will be held accountable, they will respond. But while waiting for this scenario to play itself out, valuable time will be lost, and the planet will continue to warm.[12]

More encouraging is the unprecedented attention[13] being given to climate change among those vying for the 2020 Democratic presidential nomination. This is indeed good news. We offer the following suggestions to the candidates.

Be ambitious, but not at the expense of credibility. Avoid offering up obviously unrealistic emission reduction timetables. It will only provide the naysayers with an opportunity for further procrastination as they challenge the practicality of the proposals.

Acknowledge that the carbon-producing and carbon-using capital stocks (powerplants, transport, buildings) cannot be changed overnight. Analysis of what constitutes a realistic pace is required, not just wishful thinking.

Also, identify the losers in the transition away from fossil fuels—for example, coal miners and the municipalities and states in which they live. Describe programs that will aid those who cannot help themselves. This is not only the right thing to do, but it will also be politically expedient, removing barriers to moving forward.

Finally, a suggestion to Green New dealers: Navigating our climate future will be hard enough. Lumping it together with a long list of other societal objectives will only serve to enhance political gridlock. Refocus the conversation on the other Roosevelt—specifically Teddy Roosevelt's "square deal."[14] That is,

---

[9] https://www.climate.gov/news-features/featured-images/national-climate-assessment-states-and-cities-are-already-reducing.

[10] https://www.google.com/search?ei=dKwwXZeuG8u7ggf68La4Bg&q=comapnies+tht+reduce+carbon+emissions&oq=comapnies+tht+reduce+carbon+emissions&gs_l=psy-ab.3..0i13.23860.25346..25431...0.0..0.68.818.13......0....1..gws-wiz.......0i71j0i7i30j0i8i7i30j0i7i5i30j0i13i5i30j0i8i13i30.dcB9MS5haYM&ved=0ahUKEwiXuM-b_77jAhXLneAKHXq4DWcQ4dUDCAo&uact=5.

[11] https://www.edf.org/climate/how-climate-change-plunders-planet.

[12] https://www.businessinsider.com/paris-climate-change-limits-100-years-2017-6.

[13] https://www.motherjones.com/politics/2019/07/nearly-every-democratic-presidential-candidate-is-now-backing-a-debate-on-climate-change/.

[14] https://www.texasgateway.org/resource/roosevelt%25E2%2580%2599s-square-deal.

safeguard our natural resources, particularly the ecosystems upon which our prosperity depends. Preserve capitalism, not capitalism run amuck. And seek consumer protection, especially shielding lower income groups from an unfair share of the burden.

Afterword

One of the challenges when trying to communicate with the general public about an issue like climate change is that you rarely know if anyone is listening and paying attention. This is also true about advice to politicians, and it was not evident that our suggestions had any influence as the Democratic candidates moved through primary debate season. Very few were spending much time talking about climate, perhaps not because they were unconvinced of the severity of the problem, but instead because most were on the same page. It wasn't a "bright dividing line" issue, likely to reveal sharp differences in the positions of the candidates. Later, in the general election, during the debates between the chosen Democrat and Republican candidates, the issues we addressed in this essay seemed rather subtle. The issue in the debates between Biden and Trump was whether there was even a problem requiring urgent action.

Essay 3. A Tragic Misperception About Climate Change

Prologue

In 2019, we were living under an Administration constantly repeating the mantra, "America First!" Two years earlier, President Trump had set the process in motion to take the U.S. out of the Paris Agreement. It would harm U.S. workers, he argued. And it was unfair because, he claimed, other countries (mainly China) might put forth less effort than the U.S. to reduce emissions of greenhouse gases. The Trump Administration followed through on this "America First!" approach to the climate problem, not just opposing U.S. action on emissions reduction, but also stopping the U.S. contribution to programs aiding efforts by developing countries to address the problem.

One can argue, we suppose, about the morality of a policy of "America First!" But if one takes that position, it is important to know what is in the national interest. On this point, the Trump Administration was wrong about the U.S. role in the international efforts to reduce greenhouse gas emissions. The Administration did not understand that rejecting our obligations under the Paris Agreement could be harmful to the U.S. Working with other countries to address the climate change problem was in our own national interest. We set out to explain why.

This essay first appeared as https://thehill.com/opinion/energy-environment/459980-a-tragic-misperception-about-climate-change.

A tragic misperception about climate change

Richard Richels, Henry Jacoby, and Gary Yohe

September 4, 2019

The Trump Administration's rejection of efforts to limit human influence on the global climate is not in the national interest. Along with other nations, the United States faces great risks from climate change.

The jury is in — only the most dyed-in-the-wool skeptics refuse to acknowledge that global warming is occurring and is human induced. Let there be no misunderstanding, the threat of severe economic and environmental damage is real and growing.[1] Even Trump's "America Firsters" should agree that we ought not risk the very environment upon which the health and prosperity of the nation depends.

The U.S. contributes to global warming not only through its own emissions of greenhouse gases but also by the effect of its behavior on the actions of other countries. The Trump Administration's failure to act ignores the fundamental nature of the problem. The burning of fossil fuels is the most important producer of carbon dioxide, which, when it becomes uniformly mixed in the atmosphere, impacts us all, regardless of the physical location of the source.[2] Three areas for action are indisputably in America's national interest when it comes to curbing global carbon pollution. They are:

- eliminating our own carbon pollution;
- encouraging others to do the same;
- helping poorer and highly vulnerable countries develop the capacity to adapt to the climate change we cannot stop, without which they will not be able to participate in the global carbon-abatement effort.

Most important is for the U.S. to get its own house in order. Early decades of industrialization required inexpensive sources of energy, and we developed coal, oil and gas resources to fill the bill.[3] Then, to add to the problem, we developed a consumer economy similarly dependent on these fuels. Only later did the full cost of fossil fuel burning become apparent, and now there is an obvious need for transition to cleaner forms of energy. It will be necessary to direct new investment toward low polluting technologies, and no doubt to subject some existing facilities to early retirement. The cost to replace our carbon-polluting capital stock will be substantial.

Technology choices, however, must reflect the true costs of energy — not just for the economy but also for public health and the environment. That would

---

[1] https://www.ipcc.ch/assessment-report/ar6/.

[2] https://www.epa.gov/.

[3] https://www.epa.gov/ghgemissions/sources-greenhouse-gas-emissions.

come from burdening each with a financial penalty that captures the resulting climate damage[4] estimated to result from the carbon emitted.

When including all costs of carbon pollution, the technology mix tilts in the direction of lower polluting technologies. The U.S. experience in addressing acid rain, with a charge on the emissions of sulfur pollution, demonstrates the value and feasibility of a more inclusive accounting system.

The U.S. must also encourage other countries to take similar action. Since carbon emissions respect no boundaries, it is clearly in our self-interest to act in a manner that encourages other countries to emit less. Inexplicably, current U.S. climate policy does just the opposite,[5] reflecting an extraordinary predisposition to ignore, misinterpret and misrepresent the data and underlying science. This behavior offers other countries a pretense for adopting a similar position and threatens to undermine the fledgling global climate effort just getting underway.

Offering developing countries incentives to forgo carbon dependence both now and in the future will be a heavy lift. If the U.S. does not reengage in the effort to halt carbon emissions, we will hardly be in a position to persuade others to do so, especially the three-quarters of the global population that are still in early stages of economic development.[6] Moreover, in addition to a demonstrated domestic commitment, and perhaps some moral suasion, developing countries will need technical and financial assistance to fuel their development ambitions with clean energy technologies.

Finally, it is in the U.S. interest to help the most vulnerable countries adapt to climate change that is already baked into the system. The amount of help required will depend on how quickly we manage to reduce global carbon emissions, and on mother nature's response to our assault on the global system.

Ironically, the impacts of climate change will be felt most severely by the countries least able to deal with them. Unless these poorer countries can address the more immediate challenges of sustaining their population while maintaining political stability, they will be unlikely to participate in the global carbon-abatement effort.[7] Ignoring the needs of these countries risks creating hundreds of millions of environmental refugees and stateless peoples. Regional conflicts and humanitarian crises would result — events that would tax the resources of rich and poor nations alike.[8]

The U.S. is at a critical juncture. We can remain a spectator and simply stand by and watch as damages continue to mount. Or we can rejoin the global effort to fight climate change and provide support and leadership in the struggle ahead.

---

[4] https://www.nature.com/articles/d41586-017-07510-3.

[5] https://www.ucsusa.org/our-work/center-science-and-democra/promoting-scientific-integrity/climate-change.html.

[6] https://www.un.org/development/desa/dpad/wp-content/uploads/sites/45/WESP2019_BOOK-ANNEX-en.pdf.

[7] https://www.nap.edu/catalog/12781/americas-climate-choices.

[8] https://www.ipcc.ch/site/assets/uploads/sites/2/2019/05/SR15_SPM_version_report_LR.pdf.

Committing to the global task represents the most just and ethical path; it is also the path most clearly in the nation's self-interest.

Afterword

U.S. states and cities tried to fill the gap created by the Trump Administration's rejection of the Paris Agreement. New York City mayor Bloomberg and California Governor Brown launched "America's Pledge," which reaffirmed the support of states, cities and businesses for the Paris goals. This organization sent delegations to meetings of the Conference of the Parties (COP) of the Climate Change Convention. The goal was to counteract the impression that the U.S. had completely abandoned efforts to address the problem of human-caused climate change.

After the 2020 election, the U.S. reentered the community of nations working to solve this problem. The Biden-Harris Administration rejoined the Paris Agreement. On Earth Day 2021, President Biden hosted a virtual two-day Climate Summit, including the leaders of 40 countries that account for more than 85% of the world's greenhouse gas emissions. At the 2021 COP, the Administration submitted an enhanced emissions pledge (a Nationally Determined Contribution in COP jargon). The U.S. pledged to reduce its greenhouse emissions by 2030 to 50–52% below the 2005 level.

This pledge presented the next challenge: to implement federal policies needed to meet the new goal. Developing them was going to be challenging in the face of a decision by the Supreme Court restricting the scope of administrative action on climate, and given lack of Republican support for climate action in the Congress. By 2022, it was evident that the U.S. was falling short of its targeted emissions reduction pledge. Once again, the credibility of the U.S. as a reliable partner in the global effort to tackle climate change was hanging in the balance.

Essay 4. Who Is Holding Up the War on Global Warming? You May Be Surprised

Prologue

A year before the 2020 U.S. presidential election, with the primary season hard upon us, we began to wonder why so little attention in the campaigns was devoted to climate change and programs geared towards reducing U.S. greenhouse gas emissions. Clearly, building support for such programs was challenging. A portion of the citizenry was either dismissive of the science or profoundly skeptical about whether climate change really posed a threat to their well-being. They did not see a problem and thus could not understand any need for action. However, surveys showed that this group represented only about a quarter of the population.

So what was the problem? Why was there no mention in political debate of the other three-quarters of the population, of their concerns, and of their declared support for federal action? We decided that it was time to explore these complexities.

This essay first appeared as https://thehill.com/opinion/energy-environment/468677-who-is-holding-up-the-war-on-global-warming-you-may-be-surprised.

Who is holding up the war on global warming? You may be surprised

Richard Richels, Gary Yohe, Henry Jacoby

November 2, 2019

The good news is that the American public finally appears to accept[1] that global warming is a problem. The bad news is that a substantial percentage of the public is unwilling to pay much[2] to do anything about it. At first glance these may seem to be contradictory messages. But the public may be reacting to the initial symptoms of a warming planet rather than the dire consequences envisioned by the scientific community if global warming remains unchecked.

This explanation is supported by recent findings[3] that a majority of Americans believe that the weather-related disasters we have been experiencing are becoming more severe and that the main culprit is a warmer global climate. But what the public foresees for the future is unclear. The outlook may be unambiguous to climatologists. But does the public buy into what the science shows about the implications of failure to reduce greenhouse emissions?

If the answer to this question is "no," then it may help explain why a substantial share of the public gives such low priority[4] to efforts to address longer-term climate change risk. Many people simply do not yet believe that continued procrastination will likely have catastrophic consequences for society and the environment. Perhaps a well-paid opposition has been more successful in sowing doubt than we had feared.

There are, of course, several factors contributing to the current intransigence: A belief either that a technological "fix" will save the day[5] or that government will impose the costs on someone else. Both explanations involve a lot of wishful thinking, fueled by a lack of understanding about the inertia in the physical, technological and political-economic systems.

But in any event, if the adage "to see is to believe" plays a dominant role in shaping public attitudes, we are in trouble. Due to lags in the climate system, it will take decades for many of the effects of today's emissions to play themselves out. By then, we will likely have committed the planet to much of the damage we fear the most.

---

[1] http://climatecommunication.yale.edu/publications/climate-change-in-the-american-mind-april-2019/.

[2] http://www.apnorc.org/%2520/Pages/Is-the-Public-Willing-to-Pay-to-Help-Fix-Climate-Change-.aspx.

[3] http://www.apnorc.org/news-media/Pages/AP-NORC-Poll-Most-Americans-see-weather-disasters-worsening.aspx.

[4] https://www.pewresearch.org/fact-tank/2019/02/05/republicans-and-democrats-have-grown-further-apart-on-what-the-nations-top-priorities-should-be/.

[5] https://www.researchgate.net/publication/317377828_Lay_Perceptions_of_How_Long_Air_Pollution_and_Carbon_Dioxide_Remain_in_the_Atmosphere.

Most troublesome is that, if the public is fixated on what they can see on a given day, season or year, they will be vulnerable to the machinations of those who see cold snaps as confirming that global warming is a ruse. They argue that short-term deviations are explained by the natural variability in local weather.

For example, a U.S. senator once brought a snowball[6] on to the Senate floor as proof that climate change is a hoax. That year (2015) turned out to be the hottest in recorded history until that time.

So, what has the public seen to date? The government provides an exhaustive accounting of deaths, direct economic losses and other impacts for natural disasters whose frequency and intensity are associated with climate warming. Those disasters include heat waves, severe storms, hurricanes, droughts, floods, wildfires, famines and sea level rise. Accounts of such events are also increasingly reaching the public eye, either when people look out their kitchen windows or when they turn on the evening news. What is stunning is how fast damages have risen[7] over the past four decades.

What they cannot see, however, is the relentless and mounting toll if procrastination continues. Unfortunately, this information seldom escapes the scientists' laboratory, and hence, reports of their findings fail to penetrate the public consciousness. But the day of reckoning cannot be forestalled forever. The clock is ticking.

So what can we do? Much has been written about the need for better communication and better education. Those are no-brainers. But there is other work to be done, including addressing this fundamental question: What is driving current public attitudes about climate change? That's where we need to focus more of our resources. Good natural science is critical, but so is research into the behavioral science behind the public's attitudes.

Public opinion isn't the only barrier to action. Lawmakers need to play a far greater role in combatting this existential challenge. They naturally carefully judge the mood of the public, with eyes on polls that reflect their electability. When a sufficient fraction of their constituents tilt towards action, they will be happy to jump to the front of the parade. Hopefully, when that finally happens it will not be too late.

Afterword

There is another puzzle about public support for federal climate policy that was not explored in this essay. Why are strong measures to reduce greenhouse gases implemented in so many states when effective national policies are absent? Dozens of states have implemented climate-motivated policies,

---

[6] https://www.motherjones.com/environment/2018/02/3-years-ago-we-all-laughed-at-james-inhofes-snowball-the-joke-was-on-us/.

[7] https://www.climate.gov/news-features/blogs/beyond-data/2018s-billion-dollar-disasters-context.

including regulation in transport, electric power, and building energy use. And several states have implemented cap-and-trade or greenhouse gas tax systems.

The correlation is far from perfect, but this state-level activity is found mainly in Blue (Democratic leaning) states. There is much less activity in the Red (Republican) ones, which also tend to be the origin of lawsuits intended to block federal climate measures. One possible conclusion from this observation is that even people who worry about the climate threat tend to revert to their tribal loyalty when it comes to having the federal government do anything about it. Gaining the needed U.S. action on climate, it appears, will require somehow finding a way to raise climate concerns above the deep division in U.S. society.

Essay 5. There Is No Plan B on Climate Change

Prologue

In the late fall of 2019, we had good reason to feel mounting frustration with the advancement, or rather the lack thereof, of action on climate change. A recently completed meeting of the Conference of Parties (COP) to the Climate Convention in Madrid had once again been unable to resolve long-standing disputes over details of the 2015 Paris Climate Agreement. The Trump Administration had denied climate change was a problem. Climate issues seemed to be getting little attention in the early stages of the U.S. political campaign.

There are two ways that the climate change threat could rise in importance on the long list of public priorities. One way was better understanding and widespread acceptance of the underlying science, and of the likelihood of rising climate damages established by the science. Better understanding would lead to support for action to manage the risk. Alternately, society could wait for mounting numbers of climate-augmented disasters to be convinced. The second path was going to be very costly. We used the "No Plan B" metaphor to argue for enhanced and more effective efforts to help the public to understand this choice.

This essay first appeared as https://thehill.com/opinion/energy-environment/474994-there-is-no-plan-b-on-climate-change?rnd=1576622411.

Essay 5. There is no Plan B on climate change

Richard Richels, Gary Yohe, and Henry Jacoby

December 20, 2019

It has become fashionable for critics of the Paris Climate Accord[1] to ask "What is Plan B?" That's particularly the case in the wake of failures of agreement in its just completed summit in Madrid.[2] This question is shorthand for "If the current accords prove to be unachievable, how do we plan to put the planet on a more stable footing?"

Alas, there is no Plan B. But here are some ways to move the current plan along.

The overarching goal adopted in Paris in 2015 was to hold temperature increases to 1.5 to 2 degrees Celsius relative to preindustrial levels. This will require driving net greenhouse gas (GHG) emissions to zero with progressively more ambitious commitments.

From the very beginning there has been contentious debate over whether these limits are aspirational or to be taken literally. But that's immaterial. The experience of the last four decades suggest that the climate system is becoming increasingly inhospitable to humankind — and that much of it is of our own doing.[3] Common sense suggests that we do what we can to halt emissions as expediently as possible.

Although abatement efforts have reduced global GHG emissions below earlier projections, they are still rising. To stop warming at levels even close to those suggested will require that we soon bend the curve more dramatically downward. This will not happen unless the public realizes that global warming poses an unacceptable risk to human health, economic prosperity and the very ecosystems upon which we depend. And be in no doubt, it does.

This low public support[4] for climate action will ultimately be corrected in one of two ways. The nations can let Mother Nature take the lead and allow conditions to continue to deteriorate until the public demands action. Or they can double down on efforts to inform the public what is at stake and what will be necessary to make the risks more tolerable.

Relying on Mother Nature assumes that "seeing will be believing" — that is, that substantial progress will be made only once the type and extent of the threats is painfully revealed in day-to-day climate damage. Valuable time will be squandered, and the price of the further procrastination will be heavy,

---

[1] https://unfccc.int/process-and-meetings/the-paris-agreement/what-is-the-paris-agreement.

[2] https://unfccc.int/cop25.

[3] https://www.ipcc.ch/sr15/.

[4] https://thehill.com/hilltv/what-americas-thinking/397927-pollster-climate-change-is-not-a-priority-issue-for-voters.

because the process is not reversible on human time scales. Continuing to play chicken with Mother Nature is not only irresponsible, it is utterly crazy.

The alternative is to increase efforts to heighten public awareness of the hard-won insights from the natural sciences.[5] We are not referring to the protesters in the conference halls of United Nations meetings, but those who seem blissfully unaware of the seriousness of the impending dangers. By helping the poorly-informed better appreciate what is ahead, the severest damages may be forestalled, if not avoided altogether. At the very least, we may buy valuable time to enhance our ability to adapt.

First, of course, it is important to better understand why those who deny that climate change is occurring, or that human activity is a main cause, do so in the face of overwhelming scientific evidence. Or why those who accept that climate change represents a serious risk put such a low priority on efforts to ameliorate its impacts. In short, what is driving public attitudes about climate change?

One needs only watch National Geographic documentaries or BBC specials[6] to see how well the story can be told. Unfortunately, much of the intended audience is either watching something else or somehow not absorbing the message. If communication of the risk is to be more effective, likely more guidance is needed from research by social and behavioral scientists.

More investment is also needed in improving environmental literacy. A prime candidate would be to introduce environmental science courses throughout primary and secondary school curricula. Such an effort would both improve the nonscientist's understanding of the world in which we live, as well as produce a better-informed electorate. But the clock is ticking.

Publics in many countries may be grossly misinformed about the stakes, oblivious to the science and apparently apathetic to future generations. But humankind is not suicidal. There will come a point where the costs of inaction overwhelm the status quo, and action will be universally demanded.

Whatever we can do to speed up the learning process will be effort well spent. The science already provides a compelling case for action, but it must be more effectively communicated in the court of public opinion.

---

[5] https://www.ipcc.ch/report/ar5/wg1/.

[6] https://www.bbc.co.uk/programmes/m00049b1.

Afterword

The 2020 COP was canceled because of the pandemic. All of its activity was postponed to Glasgow in 2021. In Glasgow, the main disagreements over Paris Agreement details were resolved, though the controversy over Glasgow's article on Loss and Damage remains a nagging source of frustration among developing countries. Many countries submitted revisions to their 2015 pledges that represent greater ambition, though most countries have not adopted policies that will put them on a path to meeting the new pledges. The U.S. is prominent among those with misalignment between ambitions and policies.

Over the years following this essay, climate change has risen in public attention and political debate—perhaps because of greater understanding and acceptance of the science, but surely also in response to mounting heat emergencies, wildfires, and floods. The passionate cry that "THERE'S NO PLAN B!" is worth repeating, accompanied by increased efforts to help the public understand what is at stake.

Essay 6. Adapt, Abate, or Suffer—Lessons from Hurricane Dorian

Prologue

To generate support for climate action, it is crucial to get people to look past their own kitchen windows and observe the impacts of climate change—not only in their towns, but also throughout their regions, across their nation, and around the globe. Having that broader context really matters. Hurricane Dorian's devastation of the Bahamas provided us with the chance to emphasize that broader context. We could illustrate the ever-growing risks of gradually warming the "breeding grounds" for hurricanes and show how to think about the choice of appropriate response as a problem of risk management.

This essay first appeared as https://theglobepost.com/2019/10/22/lessons-hurricane-dorian/.

© The Author(s), under exclusive license to Springer Nature Switzerland AG 2023
G. Yohe et al., *Responding to the Climate Threat*,
https://doi.org/10.1007/978-3-030-96372-9_6

Adapt, abate, or suffer – lessons from Hurricane Dorian

Gary Yohe, Richard Richels, and Henry Jacoby

October 23, 2019

There is no way to minimize the incomprehensible misery and destruction that hurricane Dorian inflicted on the Bahamas.[1] For nearly two days, anguished inhabitants were subjected to unfathomable suffering and heartbreak from a Category 5 hurricane that would just not go away.

Putting their lives back together will be the beginning of recovery after years of grief and demolition. The international effort to provide humanitarian assistance is just getting underway, and the needs are enormous – both in the immediate term and perhaps over the next decade.

Dorian is just one example of extremely damaging events whose intensity and frequency can be attributed in large measure to a human-induced changing climate.[2] Indeed, its exaggerated strength in the middle of a very active hurricane season was a well-known signature of this link, but its stalling over the Bahamas is evidence of a new signature.

Climate Change and Hurricanes

Although it is difficult to tie any single storm to global warming,[3] correlations between trends in temperature and the character of hurricanes have been evident for some time. Now it is becoming clear that climate change is altering the behaviors of atmospheric systems that steer storms of all sizes. They are showing a tendency to weaken at very inopportune times. This means that the hurricanes and cyclones that are growing in intensity due to historically hot oceans and air temperatures are now frequently being deprived of any sense of direction.[4]

New science explains why tropical cyclones around the world have, alarmingly, become prone to just sitting still for a while until conditions change. Hurricane Dorian was not the first storm to show this behavior. Harvey (2017)[5] stalled over Houston, and Florence (2017) stalled over the Carolina coastline. Dorian joined a growing and dangerous club of enormous storms that sometimes get lost. Since then, typhoon Hagibis[6] (the worst storm in Tokyo in 60 years) joined the club.

---

[1] https://theglobepost.com/2019/10/09/bahamas-dorian-clean-up/.

[2] https://theglobepost.com/2019/09/09/climate-change-human-rights/.

[3] https://theglobepost.com/2019/09/04/scientists-climate-breakdown/.

[4] https://climatefeedback.org/evaluation/national-geographic-hurricanes-moving-slower-global-warming-craig-welch/.

[5] https://theglobepost.com/2017/08/29/houston-harvey-storm-rain/.

[6] https://www.independent.co.uk/news/world/asia/japan-typhoon-hagibis-storm-destruction-fukushima-a9163101.html.

As a result, storms' destructive potentials derived from increased storm surge, elevated rainfall totals, and intensified winds, wind gusts, and tornadoes are becoming an even greater source of enormous risk. The damages from stalled tropical cyclones have simply exploded.

These new insights add renewed urgency to confronting two fundamental questions: how should we respond to growing climate risks, and are we prepared for the increasing risks?

Responding to Growing Climate Risks

In 2007, the U.N.'s Intergovernmental Panel on Climate Change concluded by consensus of over 280 nations that[7] reacting to climate change "involves an iterative risk management process that includes both adaptation and mitigation."

"Iterative" means that we should expect to make adjustments as an uncertain future unfolds and new science emerges. "Risk" is the product of the likelihood of an event and the consequences that it would impose on natural and social systems. "Risk management" is not a foreign concept; its application tells us to take account of more than just the most likely future. We do it all the time in our personal lives: we install locks for our cars and houses, invite guard dogs into our lives, and buy insurance. In responding to climate change, the same concept is being applied on a national and international scale.

"Adaptation" means working to minimize consequences. Mitigation implies reducing the strength of the sources of these consequences: abating global warming by reducing emissions.

Abatement (mitigation) is essential, but it must be recognized that the climate problem cannot "be fixed" or "rolled back." We are already committed to warming of between 1 and 1.5 degrees Celsius above pre-industrial levels. Even if we were to stop emitting polluting gases today, the pent-up heat in the oceans would put us closer to the upper part of this range. Achieving a 1.5-degree limit is therefore virtually impossible. A 2-degree limit is wildly aspirational, but even a 3-degree limit will be difficult to achieve.[8]

And so, we are left with a combination of three choices to consider: to adapt, abate, or suffer.

In making investments in adaptation, resources are critical and in short supply. In the case of Dorian, there was enough time to get people in the United States out of harm's way, but options designed to preserve residential and commercial property, protect infrastructure, or save ecosystems were alarmingly few. Sadly, by way of contrast, residents of the Bahamian Islands had nowhere to go and their properties were not nearly as robust, and so, they suffered the greatest damages.

---

[7] https://www.ipcc.ch/site/assets/uploads/2018/02/ar4_syr_full_report.pdf.

[8] https://theglobepost.com/2018/12/18/cop24-climate-change/.

Are We Prepared?

Coastal communities[9] around the world are getting an idea of what the future might look like, and it is scary. Risk management tells these communities that they may have to be satisfied with getting people out of danger as efficiently as possible when needed. They should also begin to put longer-term measures into place: productive investments that would limit the storm surges, improve road networks for escape, strengthen building standards, reduce power system vulnerability,[10] and other physical versions of insurance. These investments take time and require resources.

In the longer term, planners should consider investments whose streams of benefits accumulate over many decades. Governments of all scales must recognize that climate risk exists[11] so that they can leverage private investments, federal governments must prepare for hazards that span multiple states and provide aid for jurisdictions that are resource-starved.

In all cases, planners, governments, and citizens must prepare to handle the residual effects of possible futures that could overwhelm their plans, and all must recognize that the suffering caused by those residuals could be immense.

Risk Management

"Adapt, abate, or suffer are our *only* choices," said scientist John Holdren multiple times during his eight-year tenure as President Barack Obama's senior advisor on science and technology issues. He was right.

Since we live on a planet where climate plays an enormous role in determining the frequencies and intensities of extreme events that can affect our lives,[12] we must strive to invest in projects, policies, and technologies that make the worst futures less likely and the best futures more likely. We need to devote significant resources to plans, programs, and technologies that can abate the pace of climate change by reducing emissions even as we invest significantly in adaptation.

That is what risk management looks like at a national or global scale.

Afterword

As the *My Fair Lady* lyrics go, "In Hertford, Hereford, and Hampshire, hurricanes hardly ever happen." Perhaps not, but with warming oceans, areas affected by hurricanes are changing. And hurricanes are becoming more ferocious than Professor Higgins might have ever presumed. It is now widely

---

9 https://theglobepost.com/2018/09/21/trump-fema-hurricane-maria/.

10 https://theglobepost.com/2017/10/03/puerto-rico-solar-energy-maria/.

11 https://theglobepost.com/2019/09/09/climate-change-human-rights/.

12 https://theglobepost.com/2019/08/09/climate-change-food-security/.

accepted that these and other hurricane characteristics have been changing because of climate change. Hurricanes may not be more frequent, but they are now more intense and more damaging than they were a few decades ago.

So what lessons have been learned? The learning process is slow, but many communities affected by hurricanes and other extreme events are beginning to respond. They have begun to realize that they must consider extreme and hitherto implausible events when making long-term plans or conducting stress tests of critical public systems.

In "The Tempest," William Shakespeare wrote that "what's past is prologue" (to the future). In the time of human-caused climate change, this is no longer true. The climate of the last several millennia is not a reliable guide to the climate we will experience over the twenty-first century.

Essay 7. The Trump Administration Cooks the Climate Change Numbers Once Again

Prologue

In 2020, the Trump Administration published a new estimate of the social cost of carbon (SCC). The new number was dramatically different from that developed under the Obama Administration, slashing the old number by around 90%. We wanted to take on this topic because several of us had been involved in formulating the concepts behind SCC estimates. We had also contributed to studies of ways to improve the underlying methods and data sources.

The SCC, and equivalent estimates for methane and other greenhouse gases, are important because their main use is in benefit–cost analysis of measures that lower or raise emissions of these gases. Such analyses are required by the U.S. federal government and several states. Cutting the SCC by 90% largely wipes out the economic justification for many climate policies, which of course was the purpose of the revision.

This article first appeared as https://thehill.com/opinion/energy-environment/507929-the-trump-administration-cooks-the-climate-change-numbers-once#bottom-story-socials.

The Trump Administration cooks the climate change numbers once again

Richard Richels, Henry Jacoby, Gary Yohe, and Benjamin Santer

July 19, 2020

In its campaign against action on greenhouse gas emissions, one of the more subtle moves by the Trump Administration is its manipulation of the Social Cost of Carbon (SCC). This number is used to represent the damage resulting from emitting an additional ton of carbon.[1] Climate economists sometimes refer to it[2] as the most important number you've never heard of. Undermine the SCC and you can discredit action to fight climate change, boost support for the fossil fuel industry, tip the scales away from renewable energy and counter other important policy initiatives. Fortunately, in a detailed report on the estimation of the SCC, the congressional watchdog General Accounting Office has called out[3] this latest affront to reliable assessment of the science and risks of climate change.

The SCC is a key input to the benefit-cost analyses required of all federal regulatory actions,[4] and thus is an important factor in their justification. The federal SCC estimate has also been adopted by several states. Examples of the SCC's use are abundant,[5] including the setting of reasonable federal standards for the performance of private automobiles and appliances.

Estimating the SCC requires joint consideration of natural and social science aspects of the climate change problem. A federal working group spent nearly a decade on this process. Recognizing that the underlying methodology needed rigorous and impartial review, the interagency group commissioned a comprehensive update by the U.S. National Academy of Sciences (NAS). The 2017 NAS report supported the previous approach to valuing the SCC, recommending[6] a program of research and analysis to improve the estimate.

The Trump Administration did not follow this recommendation. Instead, it imposed measures to hobble reliable estimation of the SCC. The earlier working group was disbanded, associated documents were withdrawn and the NAS study was ignored. Instead, changes were made to limit the SCC's scope and the weight it gave to future generations.[7] These changes cannot be justified by either the science or the standards deemed acceptable for benefit-cost studies.

---

[1] https://media.rff.org/documents/SCC_Explainer.pdf.

[2] https://yaleclimateconnections.org/2015/02/understanding-the-social-cost-of-carbon-and-connecting-it-to-our-lives/.

[3] https://www.gao.gov/assets/710/707776.pdf.

[4] https://obamawhitehouse.archives.gov/omb/circulars_a004_a-4/.

[5] https://www.nhtsa.gov/laws-regulations/corporate-average-fuel-economy.

[6] https://www.nap.edu/catalog/24651/valuing-climate-damages-updating-estimation-of-the-social-cos t-of.

[7] https://www.federalregister.gov/documents/2017/03/31/2017-06576/promoting-energy-independe nce-and-economic-growth.

As a result of the Administration's changes, the previous central value for the SCC – roughly $50 per ton of CO_2 – was reduced by nearly 90 percent.

These changes are misguided and pernicious. They limit damages to those occurring within U.S. borders, and thus reflect a tragic misunderstanding[8] about climate change and the U.S. national interest. CO_2 emissions, primarily from the burning of fossil fuels, impact every person on the planet, regardless of the geographical location of the source. To limit current and future climate change damages, it is in the U.S. national interest not only to reduce its own emissions, but also to encourage other countries to do the same. The Administration's near-zero SCC does just the opposite, offering other countries a pretense for adopting positions that mimic those of the world's second-largest emitter.

There are many other causes for concern. The impacts of our emissions will be felt most cruelly by the most vulnerable Americans, and by those countries least able to cope with the ensuing damages. Ignoring the needs of these individuals and countries threatens to exacerbate societal inequities at home and to create millions of environmental refugees abroad. Humanitarian crises that would burden rich and poor nations alike are the obvious consequences. Preventing these crises is both the right thing to do and in our own self-interest.

Another critical aspect of the SCC calculation is the value placed on future generations. Intergenerational equity is a contentious topic. There are reasonable debates among social scientists about what constitutes fairness in the treatment of unborn generations. Despite these disagreements, there is convergence among scholars as to what represents a plausible range of discount factors. The Administration, ignoring the prudent advice of the NAS authors and other knowledgeable experts, provides no analysis of its own. It simply mandates a set of discount rates at the higher end of the spectrum, to the disadvantage of future generations.

In its assessment of the Administration's SCC procedure, the GAO uses careful diplomatic language. It writes that, "... the federal government may not be well positioned to ensure agencies' future regulatory analyses are using the best available science." Our interpretation is more direct: Ignoring the science to cook the numbers discredits the federal process[9] for public decision-making.

The GAO recommends that a federal agency should be made responsible for addressing the NAS report, and for ensuring that the best-available science is used in calculating the SCC. Sadly, there is little expectation that this recommendation will be heeded by an Administration that denies the reality and seriousness of the climate threat.

---

[8] https://thehill.com/opinion/energy-environment/459980-a-tragic-misperception-about-climate-change.

[9] https://www.msn.com/en-us/news/politics/trump-administration-has-been-underestimating-costs-of-carbon-pollution-government-watchdog-finds/ar-BB16J1dg.

Afterword

In its first day in office, the Biden-Harris Administration issued an executive order that provided guidelines for a revised SCC procedure. They also appointed an Interagency Working Group on the Social Cost of Greenhouse Gases to develop a new estimate. The previous estimate prepared under the Obama Administration, adjusted for inflation, was reestablished for use while the Working Group carried out its analysis.

Headed by Louisiana, a group of Republican-led states quickly sued the Biden-Harris Administration over this new SCC estimate, arguing that it was "speculative" and that it would drive up their costs and reduce revenue. A judge from the Western District of Louisiana ruled in favor of the states in February of 2022. But three months later, the Supreme Court reversed the district judge with a one-sentence order allowing the Administration to continue using the inflation-adjusted Obama-era SCC value. It is likely this issue will return to the courts when the Working Group presents its revised estimate.

Essay 8. Climate Change Is Getting Worse, and It's Harder to Predict

Prologue

On September 15, 2020, then President Trump visited California to survey the damage wrought by California wildfires. During a roundtable discussion, California's Natural Resources Secretary pointed out that five of the largest wildfires in California history were burning as he spoke. The Secretary repeatedly pointed out that climate change was the dominant cause of the severity of the fires. Trump had a soothing response to this description of the global warming that was intensifying the wildfire problem: "It'll start getting cooler … just you watch!"

This is the sort of deliberate disinformation that polluted the well of public discourse during the Trump presidency. Trump's confident prediction of forthcoming cooling also promoted the incorrect view that natural variability is the main driver of twenty-first century changes in Earth's climate. In his flawed way of thinking, it may get hotter, but nature will cool the planet down again, implying that there's no reason to expect wildfire weather to worsen as levels of heat-trapping greenhouse gases ramp up. Our essay dispels this dangerous notion.

This article first appeared as https://www.courant.com/opinion/op-ed/hc-op-yohe-climate-pendulum-1004-20201004-5xrmc4vkdfhv7nehn7zickakke-story.html.

Climate change is getting worse, and it is harder to predict

Gary Yohe

October 4, 2020

Extreme weather events are now among the most destructive ever experienced. Climate and weather disasters in the United States since 2010 have averaged $80.2 billion in economic damage with 521 lives lost per year. But the reality is, that number is going up. Between 2015 and 2019, annual damages averaged $106.3 billion with 772 lives lost.[1]

None of this is news to people who have been impacted by extreme weather and to scientists like me who have been paying close attention to these troubling trends. For decades, variable climate trajectories have wobbled along their trendlines as atmospheric concentrations of carbon dioxide resulting from human activity have increased since 1955 to produce seven of the hottest ten years in history since 2010.

We're starting to see these impacts manifest themselves in the weather we're seeing, and the science is now telling us that something new and even more threatening is afoot. It is called "regressing toward the tails"[2] — a very technical but very important new development.

What is it? It is the increasing tendency that extreme events caused by climate change never seem to be as bad as they could have been. While that doesn't sound like cause for alarm, what it really means is that it is ever more likely that the next extreme weather event — be it fire, flood, drought or heat wave — will be even worse than anything that we have previously experienced. So as a result of climate change, the most damaging impacts we've seen so far is becoming the new normal while mother nature keeps upping the ante.

Take a look at the worst wildfire year in California's history: 2017.[3] As of December 22, 2017, 9,270 fires had burned 1,548,429 acres, but this year has topped that.[4] As of Oct. 2, more than 7,900 fires had already burned 3.6 million acres with two months left in the fire season. In fact, the largest wildfire in California history has been joined since August by the third, fourth, fifth and sixth largest of all time, and they are still burning.

During the last few weeks, the California wildfires have most frequently been caused by thousands of dry lightning strikes[5] born of monsoon winds from Tropical Storm Fausto off the coast of Mexico. The strikes repeatedly hit

---

[1] https://www.climate.gov/news-features/blogs/beyond-data/2010-2019-landmark-decade-us-billion-dollar-weather-and-climate.

[2] https://papers.ssrn.com/sol3/papers.cfm?abstract_id=3600070.

[3] https://www.fire.ca.gov/incidents/2017/.

[4] https://www.fire.ca.gov/incidents/2020/.

[5] https://www.latimes.com/california/story/2020-08-19/destructive-bay-area-fires-fueled-by-rare-mix-of-intense-dry-lightning-and-extreme-heat.

drought-ravaged and tinder-dry forests during a near record heat wave. The northern part of the state is experiencing the perfect storm of these five conditions that all lie on the extreme tips of their independent frequency distributions.

Things are not just getting worse. They are also getting harder to control, prevent and predict. Human behavior used to be the usual cause of California fires, meaning future threats could be mitigated by taking strong proactive action against future events. But this year, residents can only retreat from the harm as quickly as possible and hope that the climate will change back. But he climate will not change back, and where can they go? As we continue to see, the COVID-19 pandemic is confounding evacuation efforts.[6]

The science community has been studying these weather events closely and has developed a growing understanding of what is really going on, but we are now challenged to communicate that knowledge clearly to the general public and lawmakers.

There are at least four problems in this regard. First, people often mistake weather for climate. Second, they really don't care whether next year will be a warmer or cooler than this year. Third, they find it difficult to internalize signal from noise along long-term trends. And fourth, they are very familiar with pendulums.

What's the problem with pendulums? They give people a false sense of predictability and a misguided sense of stability. People are used to being mesmerized by stationary ones, but the climate is dynamic. The climate pendulums are no longer returning to equilibrium at their lowest points in their arcs. Why? Because their pivot points are moving and accelerating in a dangerous direction (the long term trends) so that the swinging weights of the pendulums are blowing by their historical limits into some very dark extremes.

Afterword

Natural climate variability is real. It would happen even if there were no humans on the planet. But in the past two centuries, human-caused fossil fuel burning has dramatically altered Earth's climate. Natural fluctuations in climate are now being dwarfed by human-caused warming.

Think of the analogy of a pendulum. The current natural "swings" in climate are not on a pendulum whose cycles are anchored to a fixed pivot point. These natural swings in climate are now anchored to an ever-migrating pivot. The pivot is moving toward increasingly more dangerous territories for Earth's climate. Extreme events that were unheard of a few years ago are now

---

[6] https://en.wikipedia.org/wiki/COVID-19_pandemic_in_California.

happening before our very eyes. In the following essay, we argue that the extreme heat events of 2018, 2019, 2021, and 2022 are not big scientific surprises—on a rapidly warming planet, they are expected behavior.

Part III

Climate Change and the COVID-19 Virus

In December 2019, reports of a very contagious virus emanating out of Wuhan, China, began to circulate in the public media. Within three months, the World Health Organization (WHO) declared the novel coronavirus (COVID-19) outbreak a global pandemic. We were ramping up our essays on climate change and could not help but be struck by the similarities between the two threats to public health and well-being. Indeed, from a public policy perspective, climate change and COVID-19 are effectively inseparable if for no other reason than they compete for the same limited sources of funding. Yet the connections are far deeper.

The first two essays here concern the ways in which one of these problems influences the likelihood and seriousness of the other. One, Essay 9, identifies similarities in the barriers to effective policy and examines how measures to contain the pandemic may spill over into efforts to reduce climate change. In the second, we discuss how global warming accelerates forest decline and collapse, and so weakens barriers limiting the transfer of disease to humans from animals in the wild (Essay 10). And it explains how, once the virus genies are out of the bottle, extreme climate events make pandemics more difficult to control.

Another essay counters the distrust of science in these two areas and explains how scientists, whether climatologists or epidemiologists, use "what if" questions or counterfactuals (Essay 11). This is standard scientific procedure. It allows scientists to explore alternative scenarios in order to weigh the costs and benefits of acting too slowly versus too quickly, thus providing valuable information for policy choices. In the final essay in this section (Essay

12), we discuss a term that is likely to crop up (and be overused) in the context of both climate change and pandemics—the word "existential."

Essay 9. Can a Pandemic Aid the Fight Against Global Warming?

Prologue

While considering the similarity between the COVID and climate problems, and after observing the ambitious policy response to the pandemic, we began to wonder whether the U.S. experience of tackling COVID-19 might boost the nation's response to the climate change threat. Unfortunately, we quickly encountered common features of the two threats that suggested this short-term boost might be small. Both COVID and climate are bedeviled by divergent partisan views on the trustworthiness of science. Dispute over the appropriate policy responses tends to divide along the same political lines. Also, both problems tend to impose pain on segments of the population least able to deal with it, and the response to COVID did not seem to include improvements in the social safety net that would aid efforts to deal with the unevenly distributed effects of climate change.

On the other hand, we could see as early as May 2020 that efforts to control the virus were having positive effects likely to help in controlling greenhouse gas emissions. For example, actions to limit the spread of the virus were changing the workplace and the nature of meetings, reducing emissions from auto and air travel. And, in another sign of cooperation between solutions to these twin problems, part of the huge federal expenditure to counter a possible crash of the economy was creating unexpected opportunities to fund emissions-reducing infrastructure.

This essay first appeared as https://thehill.com/opinion/energy-environment/498145-can-a-pandemic-aid-the-fight-against-global-warming?rnd=1589658368.

Can a pandemic aid the fight against global warming?

Henry Jacoby, Richard Richels, Gary Yohe, and Benjamin Santer

May 16, 2020

One cannot help but look back nostalgically to a time when we thought the biggest crisis facing humanity would evolve over many years. How quickly concerns have changed. In a matter of months, the nation has been forced to commit to several trillion dollars in stimulus packages to keep the American economy on life support while many struggle to survive. But even under all of this stress, we cannot afford to forget the peril posed by an overheated planet. And we might ask what the pandemic portends for the longer-term threat of climate change. It is useful to start by reviewing some common misperceptions, and lessons not to be drawn from current events.

The first is the expectation that the value of science is now clear to all. Some people may assume it will play a more central role in future public health and environmental policy. Unfortunately, the evidence suggests otherwise. The Trump Administration continues to demonstrate a shocking disdain for the advice of experts.[1] Wishful thinking is often preferred over the advice of the world's leading epidemiologists, health researchers and clinicians.

A second optimistic view is that COVID-19 will heighten awareness of other risks the nation faces, and of the value of precautionary action. Unfortunately, there is no evidence that nature's current message will moderate the adminis- tration's reluctance to even recognize the climate threat, much less support development of a response. Even in the face of the impending death of thousands from the virus, federal management of the pandemic has been dominated by the divisive politics of the day and the pressures of the elec- tion cycle. The damage of climate change is simply not immediate enough to change the dismissive view of this issue, well-curated by the president and his party.[2]

Also worrying is a third misperception — that the reduction in emissions as we shelter in place is good news for the climate. Not so, unless throwing the economy under the bus is considered a viable emissions reduction strategy. The current blip in the long-term trend is but a transitory reduction and will not make a meaningful contribution[3] to meeting the long-term challenge of cutting emissions and stabilizing our climate.

Finally, the Administration touts the view that the pandemic hit when the economy was in terrific condition, evidenced by record low employment and a soaring stock market. That's nonsense. The present crisis shines an unflat- tering spotlight on a nation living well beyond its means, yet with a tattered

---

[1] https://www.nytimes.com/interactive/2020/03/18/us/trump-coronavirus-statements-timeline.html.

[2] https://thehill.com/opinion/energy-environment/468677-who-is-holding-up-the-war-on-global-war ming-you-may-be-surprised?rnd=1572714739.

[3] https://www.ipcc.ch/site/assets/uploads/sites/2/2019/05/SR15_SPM_version_report_LR.pdf.

safety net.[4] Dealing effectively with future global challenges, including the looming climate threat, will demand more responsible fiscal leadership from the Administration and Congress.

In the midst of this darkness, we should not fail to recognize that COVID-19 is forcing behavioral changes that could better position us to tackle climate change. For example, telecommuting is a no-brainer for reducing greenhouse emission. A new normal may also apply to business travel. Rather than having their employees fly off to distant destinations, companies are likely to reconsider the relative efficiency of virtual meetings. Consumer behavior is also likely to change. The ongoing online shopping trend may get a boost from the experience during the lockdown, lowering emissions from repeated short trips to stores.

And there may be other opportunities amid the economic devastation of the necessary state closures. The oncoming recession likely will prompt longer-term federal jobs legislation. A likely vehicle for this is the long-awaited infrastructure program. It surely will include improvements in the nation's roads and mass transit systems, which could and should be designed with emissions reduction in mind.[5]

Examples of other "targets of opportunity" include upgrading the nation's building stock and modernizing the electric power system. Bringing the aging power grid into the 21st century would improve customer cost and service while facilitating inputs from solar and wind sources. And perhaps one jobs program, the Administration's decision on the Keystone pipeline, needs reconsidering. We need to design the infrastructure of the future — not of the past.

The bottom line? The pandemic will not miraculously invigorate climate action. To be clear, we do not propose adding climate change to the daunting list of concerns motivating the current stimulus packages. The first order of business is surviving the coronavirus, and that leaves few resources and policy attention for anything else. But we must be mindful that other significant challenges require our attention. They must inform the new normal that we are already in the process of creating.

Afterword

Since we drafted this essay, some of the COVID-induced effects we anticipated have happened. Others have not. Many of the changes in work and associated travel seem to be here to stay. Although we saw only small government-supported investments in low carbon energy within the sudden gusher of federal COVID stimulus money, discussion of those investments

---

[4] https://www.nytimes.com/2020/03/31/us/politics/coronavirus-us-benefits.html?auth=login-email&login=email.

[5] https://thehill.com/blogs/congress-blog/economy-budget/494240-the-next-stimulus-bill-will-help-save-our-economy-it.

eventually yielded a fruitful outcome—substantial investments in low carbon energy are now part of the August 2022 Inflation Reduction Act.

Over time, the two threats have become less similar, for one key reason: the miracles of modern biological and medical science. Vaccines and anti-viral medicines have dramatically reduced the mortality associated with COVID, but there is no parallel "quick fix" for climate change.

One similarity remains, however: distrust of the basic science. Vaccines are the medical version of what is called "mitigation" in climate discussions—they reduce the threat of bad outcomes. And climate analysts might see anti-viral drugs as analogous to "adaptation" in their domain—they reduce damage. For the effort to develop effective responses to climate change, it is not good news that many people refuse effective COVID vaccines, while others scorn proven anti-virals in favor of discredited home cures.

Essay 10. We Cannot Ignore the Links Between COVID-19 and the Warming Planet

Prologue

The previous essay explored similarities between climate change and pandemic disease and noted that the response to COVID may yield some reduction in greenhouse gas emissions. But there was much more to be understood about the direct interactions between these two global threats. The origin of many viral diseases involves initial transfer from animal hosts to humans, and there are reasons to suspect that the warming of the planet may have enhanced the conditions that lead to this "jump." Additionally, extreme events associated with climate change create social conditions that favor the spread of infectious diseases.

It was important to explore these links. They provide yet another reason to take action on heat-trapping greenhouse gases. Understanding these interaction terms also helps to anticipate needed adjustments in the measures intended to slow the spread of COVID.

This essay first appeared as https://thehill.com/opinion/energy-environment/499604-we-cannot-ignore-the-links-between-covid-19-and-the-warming-planet?rnd=1590527443.

G. Yohe et al., *Responding to the Climate Threat*, https://doi.org/10.1007/978-3-030-96372-9_10

We cannot ignore the links between COVID-19 and the warming planet

Richard Richels, Henry Jacoby, Gary Yohe and Benjamin Santer

May 27, 2020

The emergence of COVID-19 suggests that global warming may present an even graver threat to human welfare than many recognize. As indicated in the scientific literature, not only could the current warming of our planet increase the likelihood of an air-borne pandemic such as COVID-19[1]; it could also damage our health and welfare.[2]

Let us be clear: We are not talking here about future warming, which is already of great concern. We are talking about the effects of a rise of 1°C that we have already experienced. Even with such "modest" warming, a stunning barrage of extreme events have happened in recent years, many of which cannot be explained in the absence of climate change.[3]

How can climate change increase the likelihood of a pandemic? Epidemiologists have been cautioning for several years that it wasn't a matter of whether a novel coronavirus pandemic would strike, but when.[4] It is well established that viruses jumping from the wild to humans are the major source of this threat, and we know that the degradation and decline of forests weakens the natural barriers protecting us from the source of infection.[5] Shrinking forests arise not only from deforestation caused by population growth and urbanization; global warming also affects forest health.[6]

Unfortunately, the genie is out of the bottle. Although we may not be able to undo the damage caused by deforestation to date, we may be able to reduce disease transmission by restoring the natural barriers between humans and wildlife, and discouraging their further erosion. This would require unprecedented land management programs, with particular focus on halting the devastation of tropical rain forests.

And does warming also amplify a pandemic's toll on human welfare? The answer is yes, and the effect is greatest during the months when extreme climate-related events tend to be most widespread.

Imagine what would happen if, in the midst of a virus epidemic, we were to experience a hurricane of the ferocity of a Dorian[7] stalled directly over a major metropolitan area? Or what if we were to suffer an extended record-shattering

---

[1] https://www.nature.com/articles/nature09575?page=12#citeas.

[2] https://www.nature.com/articles/s41558-020-0804-2.

[3] https://www.climatecentral.org/gallery/graphics/2019-billion-dollar-disasters.

[4] https://www.businessinsider.com/epidemiologists-on-chances-of-future-coronavirus-outbreak-2020-3.

[5] https://www.livescience.com/27692-deforestation.html.

[6] https://www.climatecouncil.org.au/deforestation/.

[7] https://earthobservatory.nasa.gov/images/event/145539/hurricane-dorian.

heatwave, such as the one that hit Europe in the summer of 2019?[8] How would we deal with the associated blackouts and brownouts, and keep people safe from COVID-19 as they tried to find refuge at crowded parks and beaches? For those fortunate enough to survive such events, social distancing to slow the virus spread would not be an option.

These are not idle threats. As this piece was being written, India and Bangladesh, while struggling with a COVID-19 outbreak, were moving hundreds of thousands of people into crowded shelters to avoid being hit by a strong cyclone.[9] On the other side of the world, the Red Cross, in anticipation of a "busy" Atlantic hurricane season, began considering hotel rooms instead of mass shelters, and developed plans for additional screening, masks, distance between cots, cleaning and disinfecting.

Hurricanes offer but one example of how to plan for the inevitable collision of extreme climate events and novel viruses. By modifying current climate change adaptation strategies, in light of what we have learned about pandemics, we can reduce avoidable breaches in pandemic containment.

Unfortunately, if efforts to contain transmission among humans were to falter in the midst of a major hurricane, heatwave or flood, one could easily imagine a renewed series of outbreaks. Recall that COVID-19 began with a single individual and rapidly spread across the entire planet. A strategy based heavily on containment, with the inevitable risk of a lapse, can be compromised by extreme climate events.

Ultimately, we must reduce greenhouse gas emissions, the root cause of the rising number of extreme climate events. Of course, the benefits of these actions extend far beyond reducing the damage from air-borne pandemics. They reduce a wide array of additional consequences associated with a warming planet — sea level rise, droughts, famines and mosquito-borne diseases, and the direct toll of heatwaves and severe storms on human health.

Today's primary focus is on pandemics and their threat to human health and welfare. Many factors aided the emergence COVID-19. But a warming planet has accelerated the degradation of forests and played a role in spreading contagion from wildlife to humans. And the rising temperatures we've already experienced present a substantial challenge to containment.

Until recently, many believed that the really serious consequences of an overheated planet lay decades in the future. Maybe not. The inextricable connections between global warming and pandemics are ignored at our peril.

---

[8] https://www.climatesignals.org/events/european-heat-wave-july-2019.

[9] https://www.nytimes.com/section/world/asia.

Afterword

This article was published in May 2020, and the next few months provided dramatic examples of the COVID-climate connection. What followed was an active Atlantic hurricane season, the most damaging for the U.S. being Hurricane Laura, which made landfall in Louisiana and Texas. Also, 2020 set a U.S. record in terms of the area burned by wildfires. Self-quarantining was not an option for the thousands caught in these disasters. They crowded into relief centers that offered ideal environments for spreading the virus. As warming continues, the already complicated health choices faced by individuals and by public officials managing disaster response will not become any easier.

Essay 11. Counterfactual Experiments Are Crucial But Easy to Misunderstand

Prologue

We were struck, early in the spring of 2020, by the modeling experiments that were being conducted by epidemiologists studying the spread of the novel coronavirus. They asked questions like: "What might have happened if we had acted one week earlier? Or two weeks earlier? What might happen if we do nothing for the next four weeks?" They expressed their results in terms of lives lost.

These were very familiar thought exercises for us. For example, the World Climate Research Program routinely asks modeling groups around the world to perform standard "benchmark" simulations—probing different aspects of climate model performance. Some of these simulations involve running the same model with and without human influences on climate. The "no human influence" runs provide a counterfactual—an estimate of how Earth's climate might have evolved in the absence of human-caused changes in greenhouse gases, particulate pollution, or land surface properties.

The point of these types of climate experiments is to enhance scientific understanding. Why did different models portray different outcomes? How large was the climate change "signal"—the difference in climate between simulations with and without human intervention? We wanted to point out that such counterfactual studies are standard scientific practice. They are not political "hit jobs," which is how former President Trump referred to epidemiological counterfactuals.

This essay first appeared as https://www.scientificamerican.com/article/counterfactual-experiments-are-crucial-but-easy-to-misunderstand/.

© The Author(s), under exclusive license to Springer Nature Switzerland AG 2023
G. Yohe et al., *Responding to the Climate Threat*,
https://doi.org/10.1007/978-3-030-96372-9_11

Counterfactual experiments are crucial but easy to misunderstand

Gary Yohe, Benjamin Santer, Henry Jacoby and Richard Richels

July 10, 2020

Between us, we have more than a century of experience in climate research, literature assessment, and scholarly support for domestic and international efforts to respond to environmental challenges. We have learned the value of rigorous scientific research, even when it challenges conventional wisdom, and of skepticism where it is appropriate. As we watch the response of epidemiologists and public health experts to COVID-19 (e.g., here,[1] here,[2] here[3] and here[4]), we have some idea of the challenges they face. We have seen this play before. Scientists respond to a need to provide information that will save lives; scientists are subjected to political attack for their efforts. We don't know how the play ends—but we've seen enough to know what happened in the second act of the climate change version.

Efforts to understand the behavior of COVID-19 and estimate its future spread began early in 2020. As part of this effort, researchers at Columbia University conducted a counterfactual exercise[5] to answer an important question: What would have happened if nontherapeutic interventions in the United States had started before March 15? According to their calculations, starting only a week earlier, on March 8, could have saved approximately 35,000 U.S. lives and avoided more than 700,000 COVID-19 cases through May 3 (a 55 percent reduction from what happened). Starting interventions another week earlier could have reduced deaths by more than 50,000.

On June 8, *Nature* published two more counterfactual studies. Solomon Hsiang and colleagues[6] focused on six countries (China, France, Iran, Italy, South Korea and the U.S.) that had imposed travel restrictions, social distancing, event cancellations and lockdown orders. Their calculations, supported by an estimate that COVID-19 cases had doubled roughly every two days starting in mid-January, suggested that as many as 62 million confirmed cases (385,000 in the U.S.) had been prevented or delayed through the first week in April.

In the second *Nature* study,[7] Seth Flaxman led a group that focused on 11 European countries. They worked with estimated viral "reproduction rates" between three and five; that is, every infected person was expected to infect between three and five other people per unit of time. This number, called the

[1] https://www.sciencemag.org/news/2020/02/scientists-are-racing-model-next-moves-coronavirus-thats-still-hard-predict.

[2] https://medicalxpress.com/news/2020-03-covid-impact-health.html.

[3] https://www.imperial.ac.uk/mrc-global-infectious-disease-analysis/covid-19/covid-19-reports/.

[4] https://www.policymed.com/2020/03/ihme-creates-covid-19-projection-tool.html.

[5] https://www.medrxiv.org/content/10.1101/2020.05.15.20103655v1.full.pdf.

[6] https://www.nature.com/articles/s41586-020-2404-8.

[7] https://www.doi.org/10.1038/s41586-020-2405-7.

"serial interval," is estimated for COVID-19 to be roughly four days.[8] Flaxman and his colleagues calculated that 3.1 million deaths (plus or minus 350,000) were avoided through the end of April, but they found that only lockdowns produced statistically significant effects on the number of estimated cases.

Are these high numbers really physically plausible? Yes. The virus is virulent and exponential growth is powerful. Left to its own devices, COVID-19 reproduction in humans increased at a daily rate of nearly 34 percent over the study period. If you were 20 years old and could find a tax-exempt asset that would pay that as an annual return for the next 44 years, then a $1 investment today would allow you to retire with a $3.1 million nest egg at age 65.

All of these results must be judged in their complete and proper contexts. They describe alternative assumptions about the form and timing of a response to COVID-19, leading to different trajectories for cases and deaths attributed to the virus. Each imagined path also involves policy interventions that have other economic and social effects.[9] Ultimately, it is up to decision-makers to consider the implicit tradeoffs between these intertwined impacts, and to make some overall assessment of joint levels of tolerable risk. This is a judgment that they cannot honestly make unless they acknowledge the veracity of what the science is telling them.

We are alarmed that the U.S. president took the Columbia analysis as a personal attack on his handling of the pandemic. "Columbia is a liberal, disgraceful institution," he asserted.[10] "It's a disgrace," he continued, "that Columbia University would do it, playing right to their little group of people that tell them what to do." The Columbia report was, according to the president, nothing more than a "political hit job."[11] We are even more dismayed that conservatives equate their feelings about coronavirus models with the "detest" that they feel about climate models.[12]

Let's return to the play we mentioned above. In climate world, the first act involved doing the science and conducting counterfactual experiments similar to those produced for the virus. Consider, for example, the finding that human activity is the primary cause of observed planetary warming since the beginning of the industrial revolution. This conclusion results from a well-defined set of counterfactual exercises, wherein ensembles of climate models were run with and without greenhouse gas emissions.[13]

In the second act of the climate play, scientists cope with public and political reactions to their findings. We know Act II well. Coronavirus modelers are now

---

[8] https://wwwnc.cdc.gov/eid/article/26/6/20-0357_article.

[9] https://www.barrons.com/articles/the-danger-of-overreliance-on-epidemiological-models-515881 79008.

[10] http://fullmeasure.news/news/full-episodes/full-measure-may-24-2020-interview-with-the-president.

[11] https://abcnews.go.com/Politics/study-finds-earlier-coronavirus-restrictions-us-saved-36k/story?id= 70808611.

[12] https://www.sciencemag.org/news/2020/04/us-conservatives-who-detest-climate-models-add-new-target-coronavirus-models.

[13] https://www.ipcc.de/.

living through it. In some countries and in many areas of science, scientific findings are generally accepted, and Act II seems implausible. But in the U.S., science has frequently been dismissed out of hand or ignored—a victim of misinformation campaigns designed by those with personal and/or institutional stakes in the results.

The COVID-19 counterfactuals were not a "disgrace" or "hit job." They are standard operating procedure—skillful applications of an investigative procedure that is one of the fundamental ways that serious science is performed.

This understanding of the role of science is why we argue that these particular counterfactual studies are so important. They provide rigorously supported insight into the human cost and benefits of decisions that were or were not implemented. The counterfactuals are lessons about the consequences of disregarding warnings that emerge from scientific analysis—including the warning conveyed[14] by the exiting Obama Administration, which, based on the best-available science, highlighted the urgency of early, decisive action in the case of a novel virus outbreak anywhere in the world.

The actual numbers of deaths and infections are not the message here. The real news is that they are big and believable, and that ignoring science can be very costly. The blockbuster corollary is that even a little bit of delay (or acceleration) in implementing decisions can matter a lot. It is a profound message that puts climate scientists in the same theater seats with the COVID scientists.

Afterword

Policy studies conducted with integrated assessment models (IAMs) are another useful application of counterfactual thinking. An IAM combines a simple model of how the climate works with a representation of the economic system. The focus of most IAM simulations is on the energy sector—the dominant source of greenhouse gas emissions. These studies begin with an analysis of a "no policy" case: the likely future state of the economy and the climate if no additional action is taken. The "no policy" projections are then compared with counterfactual "what might be" scenarios that consider the outcomes if specific climate-friendly policies were put in place. The results then include assessments of not just the climate effects, but also the cost and distributional consequences of the particular policy action.

The climatic parallel with the COVID "what if" investigations that motivated this essay are the studies of the increase in cost of achieving a temperature target—like limiting warming to less than 2 °C—if nations wait years to take action on reducing greenhouse gas emissions.

---

14 https://www.news-medical.net/news/20200515/Evidence-shows-Obama-team-left-a-pandemic-e28098game-plane28099-for-Trump-administration.aspx.

Essay 12. Climate Change and COVID-19: Understanding Existential Threats

Prologue

In 2020, a great deal of hyperbole muddied the waters of rational discussion about how to combat climate change and its impacts. One of the most questioned terms was "existential threat." Some in the environmental community seemed to imply the climate change risks included human extinction. To deniers, this was an obvious overstatement, supporting their view that climate change was a hoax. It struck us that framing the debate in terms of accepted definitions of "existential" could be helpful.

We took our motivation from the experience with the use of this term in connection with pandemic disease. For any unvaccinated person, the COVID virus poses a potentially mortal threat. But the human race will survive, as it did after the 1918–1919 Spanish flu pandemic (and without the benefit of current medical technology). Similarly, the human race will survive climate change. Many of its members and forms of civilization may not survive. Whether or not a threat is "existential" depends on the scale at which you look at it.

This essay first appeared as https://thehill.com/opinion/energy-environment/565698-climate-change-and-covid-19-understanding-existential-threats.

Climate change and COVID-19: Understanding existential threats

Gary Yohe

July 30, 2021

Some people are still falsely saying that climate change is a hoax.[1] Many are also falsely saying the same thing about the COVID-19 pandemic.[2] They are wrong on both counts. Both climate change and the variants[3] of the novel coronavirus are profoundly real.

Others, who point to the spate of enormously damaging climate-related events, say that we are in a climate "crisis." And it's not the only preventable crisis: On July 27 COVID-19 cases in the United States climbed to 108,000, the first time above 100,000 cases since Valentine's Day.[4] A "crisis" is, by definition,[5] "a time of intense difficulty, trouble or danger."

Then there are those (many of the same people who use the term "crisis"), who say that both climate change and COVID-19[6] pose "existential threats" to humankind. Is that hyperbole? It is certainly the polar opposite of claiming that either is a "hoax."

Asteroids or volcanoes or something else global (like a virus?) were clearly existential threats for dinosaurs. Warming oceans with increasingly acidic low pH levels are demonstrably existential threats to coral reefs all across the planet because they cannot migrate. But neither COVID-19 with its Delta variant and subsequent variants nor climate change with its myriad of increasingly frequent and intense impacts are existential threats to the entirety of the human race — at this point in time.

It is possible, though, to offer a more precise variant of the standard definition of an "existential threat"[7] – "A threat to something's very existence when the continued being of something is a stake or in danger." To be germane for one human being or his family in today's climate, I suggest: An external and global source of stress that creates random events which would put a randomly selected person, his or her family, or community or region or country in mortal danger if those events were to occur where he or she lives.

---

[1] https://www.americanprogress.org/issues/green/news/2021/03/30/497685/climate-deniers-117th-con gress/.

[2] https://news.yahoo.com/evangelical-pastor-demands-churchgoers-ditch-101435205.html.

[3] https://thehill.com/opinion/energy-environment/499604-we-cannot-ignore-the-links-between-covid-19-and-the-warming-planet?rnd=1590527443.

[4] https://www.nytimes.com/interactive/2021/us/covid-cases.html.

[5] https://www.lexico.com/en/definition/crisis.

[6] https://www.swissre.com/risk-knowledge/risk-perspectives-blog/covid-19-is-an-existential-threat-but-so-is-the-climate-crisis.html.

[7] https://www.realcleardefense.com/articles/2016/02/10/what_is_an_existential_threat_109009.html.

We know that COVID-19 is currently an existential threat for individuals because we know, on July 27 for example, that the Delta variant will infect five people who live somewhere on the planet in the next second. In short, COVID-19 is an existential threat to any human being, and so it is existential to us all to a degree that depends on where we live, the color of our skin, how rich we are and whether or not we are fully vaccinated.

As well, we know that climate change is currently an existential threat for anybody who lives anywhere on the planet because of fires, heat, floods, drought and other extreme weather. How so? Because climate change[8] will unnecessarily kill 85 of every 100,000 people on the planet every year – most in Africa, but generally on every continent except Antarctica. It follows that climate change threats are as existential to us all as COVID-19.

Not surprisingly given the many parallels that have already been discussed, using "existential" this way again makes COVID-19 and climate change look like the same problem on different time scales. So, how can we reduce these threats?

1. For COVID-19, we can wear a mask whenever possible to lower the likelihood of infection and take the vaccine both to lower the likelihood of infection and also eliminate the more severe consequences.
2. For climate change, we can invest in mitigation to lower emissions and thus the likelihoods of some extreme events over the long run, but we can also invest in adaptation that can directly lower consequences immediately.

As instructive as this parallel is, there is one difference. Adapting for COVID-19 eliminates the most severe consequences of contracting the virus. Adapting to climate change, on the other hand, eliminates its less severe but more likely consequences of extreme events.

For that reason, reducing emissions to lower the likelihood of the extreme consequences of climbing greenhouse gas concentrations in the atmosphere is even more critical than COVID-19 prevention measures.

Afterword

The term "existential" continues to be widely applied to the threat of climate change. Though to most people it may mean little more than "really, really serious," it remains important to have a practical and meaningful limit to the scope of the risk it describes. As suggested in the essay, pandemics and climate change are mortal threats to vulnerable human individuals and groups throughout the world, but neither threat is globally existential.

The same characterization applies to ecosystems. The earth as a whole will survive a warming climate, but a range of valued ecosystems will not. For example, many coral reefs may suffer complete extinction in response to the

---

[8] https://www.bloomberg.com/news/articles/2021-07-29/warming-planet-means-83-million-face-death-from-heat-this-century.

warming of the oceans and the acidification that results from the diminished ability of oceans to soak up human-emitted CO_2. As mountainous areas warm, there is only so far that Alpine ecosystems can follow their favored environment upward—they eventually run out of real estate. Likewise, many species dependent on polar ice will face extinction when ice disappears with melting. Thus for many local ecosystems and animal and plant species, the word "existential" is an appropriate description of the climate risk they face.

Part IV

The Yale Project for the Campaign Season

In anticipation of increased interest in climate change in the 2020 presidential campaign, we saw an opportunity to counter the false claims of climate change deniers. Our goal was to provide the public with accurate information about the urgency of the problem of human-caused climate change. The result was these eight essays, published by Yale Climate Connections. We saw the presidential debates as a chance to reveal the different positions of the candidates on the climate threat. The first essay (Essay 13) laid out several questions designed to focus the attention of the moderators and audience alike.

The next six essays methodically worked through the misleading claims of opponents to climate action. The first claim is the assertion that the climate is not changing, and that climate change is a hoax—a claim often made by candidate Trump (Essay 14). When this position becomes untenable in the face of the mounting evidence, the deniers then assert that humans are not the cause of the changing climate. Again—alas!—the evidence to the contrary is irrefutable (Essay 15).

When the "no human influence" position becomes too difficult to defend, the next fall-back position is that the impacts of human-caused climate change are not that serious, and, in any event, human societies are resilient. So, the deniers argue, there is no reason to believe a response is necessary even if we continue to warm the planet. Two of the essays (Essays 16 and 17) debunk that incorrect view.

Their next argument is that we don't know enough. Given the uncertainties, proposals to cut emissions could turn out to be too expensive. We should take a "wait and see" approach. In response, the next essay in this brief tour of the denier playbook makes the case for urgent action—balancing the risk of overreacting with that of doing too little (Essay 18).

Finally, there is the "But what about Country X?" deflection—the argument that certain other nations are not doing their part to reduce greenhouse gas emissions. The last of this Yale series of essays highlights the crucial impact of U.S. leadership in the willingness of other nations to take action (Essay 19).

The various arguments against acting in response to growing climate risks had been evolving for years. By the summer of 2020, these arguments were being pushed hard in the public sphere and we feared that key findings of climate science would be lost in the noise. We thought of our task as providing the members of the general public with the information they needed to rebut the specious arguments of climate change deniers and to elect representatives who pay attention to science. In the end, it is the voting power of the public (if that power can survive systematic assault) that will determine the national response to the climate challenge.

Essay 13. Five Science Questions That Ought to Be Asked at the Debate

Prologue

During the summer of 2020, as the U.S. national election was rapidly approaching, it became clear to us that the two presidential candidates treated empirical evidence and scientific analysis in very different ways, particularly regarding their relevance for public policy. The differences applied not only to the climate threat, but also to the response to COVID-19. One candidate argued for following the science without claiming that there were "sure fire" solutions, highlighting productive actions we could all take, as individuals and as a nation. The other candidate was dismissive of scientific knowledge, especially when it pointed the country in a direction that diverged from his personal and political interests.

We believed that the candidates' responses to these or similar questions would provide voters with useful information as they weighed their choices. We hoped our list would attract the attention of the media organizations hosting the debates and the debate moderators.

Though the questions that we proposed were formulated to draw out the differences between the candidates, we believed it was important to avoid "gotcha" questions, and to take care not to frame questions in terms that were too partisan. At the same time, we were well aware of the ability of public officials to side-step inconvenient questions, and so suggested follow-up inquiries for each question we posed.

This essay first appeared as https://yaleclimateconnections.org/2020/09/five-science-questions-that-ought-to-be-asked-at-the-debate/.

Five science questions that ought to be asked at the debate

Gary Yohe, Henry Jacoby, Benjamin Santer, and Richard Richels

September 4, 2020

In past elections, debates have largely ignored issues like climate change and global pandemics. This year, it's likely to be a different story. Americans want to know how the next President of the United States intends to protect them from the twin scourges of climate disruption and COVID-19. How well do the candidates understand these issues? Is their understanding grounded in science or in wishful thinking? What are their plans for responding to the relentless warming of the planet, and to a virus that has already taken more than 185,000 American lives, with many more coming?

This year's debates might be more informative if moderators and questioners begin to prepare such questions now. Let's phrase them carefully, think about what the candidates might say in response, and craft follow-ups. The goal is simple – to bring into the open real differences in the candidates' understandings, positions, and policies. Let's expose the bright dividing lines on these important issues, while avoiding "gotcha" questions that generate more heat than light.

Here we offer a short list of sample questions on issues related to climate, the environment, public health, and the value of science. Each question comes with a few sentences to add context and possible follow-up questions.

Question 1

Context: The United States has historically been a leader in international efforts to solve global problems like climate change, arms control, and infectious diseases. Some express concern that we have in many ways given up that role.

Question: Please give an example of a global science-based issue of such importance that you would give it your time and personal capital as president.

Follow-up: How would you organize, lead, and promote international collaboration in confronting the issue you identified?

Question 2

Context: In the United States, as in the world, low-carbon energy sources are already cost-competitive with fossil fuels, and authoritative data and forecasts see renewables as being a promising source for well-paying jobs in the future. Countries capable of providing inexpensive low-carbon energy will be the economic leaders of the 21st century.

Question: Many experts say developing and implementing renewable energy presents an enormous opportunity to grow jobs and the economy. Do you agree? If so, how would you make the U.S. a global leader in this effort?

Follow-up: Do you see a role for private-public partnerships in bringing low-carbon energy sources to market? Which federal agencies and what policy initiatives would be key in such an effort?

Question 3

Context: Many who remain in agencies like the CDC, EPA, NOAA, USGS, and NIH feel that the integrity of their science is being undermined. This is of real concern if the U.S. seeks to maintain international scientific leadership – and to attract the best and brightest students.

Question: A large number of highly specialized scientists have left the federal government, taking their knowledge and skills with them. What would you do to bring them back or attract qualified replacements?

Follow-up: Is it a priority for you to bring scientists back to a rejuvenated and welcoming federal research program? How will you do that?

Question 4

Context: Many proposals have argued that rebuilding our economy offers a historic opportunity to make us safer from the ongoing and future risks of climate disruption. These risks are real – they are already significantly affecting our homes, our lives, our livelihoods, and our health.

Question: We are now rebuilding the United States economy in the wake of the COVID-19 pandemic. Is this an opportunity to make our infrastructure and society less vulnerable to climate change?

Follow-up: Can action to reduce climate change risks help make the United States more resilient, prosperous, and secure?

Question 5

Context: The COVID-19 vaccines issue presents a concrete example of a potential anti-science bias that would have a significant impact on health and on our healthcare system. Ignoring vaccine science jeopardizes the well-being of all U.S. citizens – not just those who avoid vaccination.

Question: We soon hope to have a safe and effective COVID-19 vaccine. We know, however, that some, and perhaps even many, may refuse it for themselves and their children; they will encourage others to do the same. How would you respond to them?

Follow-up: Effective anti-COVID measures – like the simple wearing of face masks – have become politically divisive issues in the United States. What would you do to prevent similar political division in deploying an effective coronavirus vaccine?

We hope that this little exercise stimulates some thought and discussion. Ideally, it will provoke prospective moderators, journalists, and the public to think carefully about the value of science in a complex, risky world. Our personal bottom line is to help ensure that science and scientific understanding get proper consideration. Science should inform at least some of the questions posed in the upcoming debates to candidates for U.S. president and vice president. We hope that science will also inform some of the answers.

Sources used in preparing this set of recommendations:

Question 1:

* U.S.'s global reputation hits rock-bottom over Trump's coronavirus response[1]

Question 2:

* 2020 is our last, best chance to save the planet.[2]

* Science under attack: How Trump is sidelining researchers and their work[3]

Question 3:

* Perceived losses of scientific integrity under the Trump Administration: A survey of federal scientists[4]

Question 4:

* Key investments can build resilience to pandemics and climate change[5]

* Priorities for the COVID-19 economy[6]

---

[1] https://www.theguardian.com/us-news/2020/apr/12/us-global-reputation-rock-bottom-donald-trump-coronavirus.

[2] https://time.com/5864692/climate-change-defining-moment.

[3] https://www.nytimes.com/2019/12/28/climate/trump-administration-war-on-science.html.

[4] https://doi.org/10.1371/journal.pone.0231929.

[5] https://www.wri.org/blog/2020/04/coronavirus-pandemic-climate-change-investments.

[6] https://www.project-syndicate.org/commentary/covid-2020-recession-how-to-respond-by-joseph-e-stiglitz-2020-06.

Question 5:

* Understanding opposition to vaccines[7]

Afterword

Chris Wallace of Fox News moderated the first of three planned debates. It was supposed to cover six topics: the Trump and Biden records, the Supreme Court, COVID-19, the economy, race and violence in our cities, and the integrity of the election—none directly concerning the climate issue. The debate was chaotic, and of limited value in terms of conveying information about the differing positions of the candidates. Candidate Trump repeatedly interrupted his opponent and was in frequent conflict with the moderator.

The Commission on Presidential Debates canceled the second debate after candidate Trump refused to accept the virtual debate format that had been imposed in response to COVID risk. The third meeting was held as scheduled, in person, with Kristen Welker of NBC moderating. Each candidate's microphone was muted during the other side's initial responses. As in the case of the first debate, there were six topic areas. This time, one topic was climate change.

None of the questions from our Yale essay was asked directly, but much of the back and forth touched on our second question about economic cost and opportunity. Candidate Trump used economic arguments to rail against action on climate risk. Candidate Biden emphasized that a well-designed program to cut greenhouse gas emissions would favor job creation. The debate never raised the deeper issue of trust in science.

Our question on the flight of scientific talent from federal agencies was not addressed either, though it focused on an important emerging problem. After the 2020 election, federal agencies would later face severe shortages of qualified staff, constraining the new Administration's ability to pursue its climate agenda.

---

[7] https://www.healthline.com/health/vaccinations/opposition.

Essay 14. Evidence Shows Troubling Warming of the Planet

Prologue

Our first task in countering climate denial was to establish the fact that the temperature of the globe has been changing over the past century at a pace not seen for millennia. It was also (sadly) necessary to rebut the denier argument that warming is a hoax. The irony of the deniers' claim is that it is being made in the face of mounting evidence of rising global average temperatures, increasing numbers of record-breaking hot days, and extraordinarily impactful heat waves.

We began our "rebutting the deniers" essay series by explaining how scientists use observations and scientific understanding to conclude that the rise in global temperature since the mid-1800s is indisputable. There is compelling evidence of global warming from many different independent measurement platforms: surface thermometers on land; ships, buoys, drifters, and satellites for the ocean; and satellites and weather balloons for the atmosphere.

We also articulated the differences between day-to-day fluctuations in weather and longer-term changes in climate—the average weather over decades. And we laid the groundwork for explaining (in a subsequent essay) why gradual, planetary-scale global warming should be expected to change the behavior of extreme weather events.

This essay first appeared as https://yaleclimateconnections.org/2020/09/evidence-shows-the-planet-warming-on-average-at-an-increasing-rate/.

Evidence shows troubling warming of the planet

Henry Jacoby, Gary Yohe and Richard Richels

September 18, 2020

Public opinion surveying consistently indicates[1]* that large portions of the U.S. population understand that global warming is happening and occurring at a troubling pace. By a ratio of six-to-one (72% to 12%), the public accepts global warming as a reality.

Six-to-one odds like that would be widely accepted as good news in any sports contest or election. So much so that some wonder what factors are driving that dismissive 12%, or the remaining 16% who declare that they just don't know.

One contributing factor is the difficulty of understanding our very noisy planet based on an individual's personal experiences. What people see is the weather – daily temperature, rain and snowfall, and an occasional storm – all of which vary naturally day to day, season to season, year to year, and place to place.

Climate change, in contrast, involves the progressive shift in these patterns over decades, and few people can clearly remember what July was like 30 or 40 years ago. Occasionally an extreme weather event may shake attitudes about climate change, but the first step in appreciating the climate threat is acceptance of evidence based on over a century of measurements.

So what is this evidence? Several groups prepare estimates of the path of global temperature since the beginning of the industrial age. Two of these are in the U.S.: NASA's Goddard Institute for Space Studies and the National Oceanic and Atmospheric Administration (NOAA). Others include the UK Met Office's Hadley Centre, the Japan Meteorological Agency, and several non-government teams.

Each of these science-based groups combines information from numerous datasets, including air temperature (e.g., from long records of meteorological agencies), ocean surface temperature collected by ships and specialized buoys, and (since the 1970s) land and ocean temperature measures derived from data collected by satellites.

Somewhat different methods are applied in these studies as findings from different sources are combined into a global picture. But despite these differences their estimates don't vary much. Results from several of the groups[2] for the increased global average temperature over the past 140 years are shown in Fig. 1. Over this period the planet has warmed by a bit more than 1 degree Celsius (slightly more than 2 degrees Fahrenheit). But even more troubling than the 140-year change is the rapid increase over just the past 50 years.

---

[1] https://climatecommunication.yale.edu/visualizations-data/ycom-us/.

[2] https://climate.nasa.gov/news/2945/nasa-noaa-analyses-reveal-2019-second-warmest-year-on-record/.

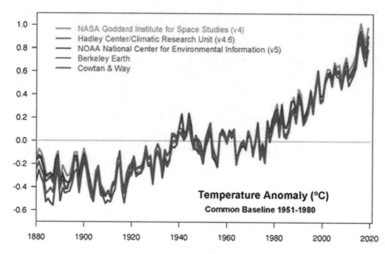

Fig. 1 Annual global average temperature, 1880–2019[3]

Every place on the globe does not warm at the same pace, of course. There is substantial variability in temperature by region over time, with the poles, particularly the North Pole, seeing the most rapid warming. That pattern is strikingly illustrated by dynamic plots of the sequence of change[4] over the period.

But even with substantial year-to-year temperature variation at a global scale, as evident in Fig. 1, five of the hottest years on record have occurred since 2015.[5] Moreover, Earth is on a path to continued warming. This last observation raises a question of what is causing this change, and the upcoming essay in this series, to be posted September 25, will show how peer-reviewed science attributes most of the observed warming to human activities that emit heat-trapping gases.

It is not only the change in average warming that is a concern, of course, but also what is happening to temperature highs and lows. In any period, say a month, the daily temperature ranges above or below the monthly average, as illustrated[6] in Fig. 2. Local weather forecasters' comparisons of the day's high or low temperature with the average for the particular time of year are staples of local TV weather reports. As the average temperature rises, the number of unusually hot days should be expected to go up, with fewer very cold days. Indeed, that is what is happening,

Consider, for instance, the U.S. experience. At each of thousands of weather stations located around the country, special note is taken when a new high

---

[3] https://earthobservatory.nasa.gov/world-of-change/global-temperatures

[4] https://climate.nasa.gov/interactives/climate-time-machine.

[5] https://yaleclimateconnections.org/2020/09/evidence-shows-the-planet-warming-on-average-at-an-increasing-rate/v.

[6] https://www.climatesignals.org/climate-signals/extreme-heat-and-heat-waves.

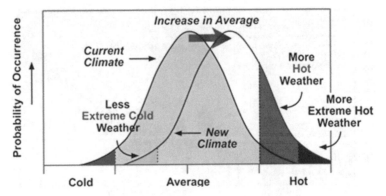

Fig. 2 The effect on extreme temperatures when the average increases[7]

or low daily temperature is recorded, and the same for a new high or low monthly average. In the past year, the number of new record monthly highs exceed new record lows by two to one.[8] The October 25 essay in this series will explore implications of a continued breaking of high-temperature records that comes with the warming in Fig. 1, as illustrated in Fig. 2.

The 2020 disastrous late-summer heat wave in the western U.S. no doubt will add to the growing collection of broken temperature records. However, individual heat events like this, though more likely with rising average temperatures, are not a good basis for swaying those dismissive of data that Earth is warming, or convincing those who are still not sure. All too soon, the planet's noisy weather system will provide a contrasting event far from the average – perhaps a Washington D.C. snowfall – to be used as evidence against warming by those determined to oppose climate action.

The key to understanding the climate change threat is in the long temperature record, requiring only confidence in thermometers – the temperature records collected by thousands and thousands of people in countries worldwide, and pulled into a global picture by several independent analysis groups.

The lead researcher for this report directs the Yale program that publishes this site.

Afterword

This essay appeared in September 2020. The climate data for the year 2020 were still incomplete at that time. When results for the full year became available[9] in January 2021, it was estimated that 2020 had the second highest

---

[7] https://19january2017snapshot.epa.gov/climate-change-science/understanding-link-between-climate-change-and-extreme-weather_.html

[8] https://www.ncdc.noaa.gov/cdo-web/datatools/records.

[9] https://berkeleyearth.org/global-temperature-report-for-2020/.

global temperature since 1850. Indeed, on land, it was the warmest year in more than 170 years.

The warming in 2020 was all the more remarkable since it was not an El Niño year. It is well established that the warmest years globally tend to occur when the natural El Niño-La Niña temperature oscillation in the Pacific Ocean is in its El Niño phase. Global surface temperature did not have this "El Niño boost" in 2020. Also, the cooler "La Niña" phase persisted into 2021. Despite this natural cooling influence, 2021 still recorded the sixth highest temperature in the entire record.

Unfortunately, despite this overwhelming scientific evidence, distrust of the findings of global-scale warming is still all too common in the public arena (see the "letters to the editor" section in your local paper).

Essay 15. The Evidence Is Compelling That Human Activity Is the Main Cause of Global Warming

Prologue

In the jargon common among climate analysts, the previous essay concerned itself with detection (the process of showing if a significant climate change is happening). Next comes attribution (explaining the cause or causes of the detected change). Responding to the deniers' fallback when they lose the detection point (their claim that warming is happening but not human-caused), we proceeded to summarize the evidence attributing significant warming to society's emissions of greenhouse gases. A massive body of scientific evidence shows that the large increase in Earth's surface temperature since the late nineteenth century cannot be explained by natural causes alone. Human influence is essential to explaining the observed warming, both in terms of its size and geographical pattern.

Attribution generally requires results from both observations and computer models of the climate system. The computer models can be used to perform the controlled "thought experiments" that we cannot perform in the real world. In computer model world, we can change a single influence at a time and see how the climate system responds. For example, modelers can increase greenhouse gases alone (according to how human activities have actually changed greenhouse gas levels in the atmosphere) and try to isolate the climate "fingerprint pattern" associated with that specific factor. Or they can run the computer model with purely natural changes in the Sun's energy

This essay first appeared as https://yaleclimateconnections.org/2020/09/the-evidence-is-compelling-on-human-activity-as-the-principal-cause-of-global-warming/.

G. Yohe et al., *Responding to the Climate Threat*, https://doi.org/10.1007/978-3-030-96372-9_15

output. This allows scientists to identify the fingerprints associated with different factors and to disentangle human and natural influences in observed climate records.

The evidence is compelling that human activity is the main cause of global warming

Gary Yohe, Henry Jacoby, Richard Richels

September 25, 2020

In our previous essay[1] in this series, we showed that the global average temperature has increased since early in the industrial revolution, rising at an accelerating pace in the past 50 years. It is no longer controversial that warming has been established: It's been proven using a long time-series of high-quality scientific data collected through well-understood measurement techniques.

So, *Why* is the planet warming? An early answer to that question dates from more than a century ago. Nobel Laureate Swedish chemist Svante Arrhenius in 1896 published an analysis concluding[2] that a doubling of CO_2 in the atmosphere would "raise the temperature of the earth's surface by several degrees Celsius." Arrhenius thought it would take several hundred years to burn enough fossil fuel to achieve that doubling, but we of course are on track to reach that troubling milestone much more quickly. The science community has very high confidence that Arrhenius was right in hypothesizing the connection of CO_2 to climate change. All these years later, his estimate of increased global temperatures resulting from more CO_2 in the atmosphere is consistent with contemporary estimates reported[3] from dozens of independent studies. These recent analyses take into account our understanding of the physics of the planet, experience in the modern era, and evidence[4] from ice cores going back hundreds of thousands of years.

So why the scientific confidence in attributing the warming to human emissions? Because an abundance of temperature data matches with high-quality *observations*[5] of CO_2 emissions and, since 1955, with the resulting atmospheric concentrations.[6] Using these data, the global scientific community has confirmed, by detailed analyses using computer models of how our planet works, that the only plausible explanation for the warming is the emissions of greenhouse gases by humans.

The point here deserves emphasis: We know from observed temperature increases (and not simply through modeling) that Earth is warming. No one

---

[1] https://yaleclimateconnections.org/2020/09/evidence-shows-the-planet-warming-on-average-at-an-inc reasing-rate/.

[2] https://agupubs.onlinelibrary.wiley.com/doi/pdf/10.1029/98EO00206.

[3] https://agupubs.onlinelibrary.wiley.com/doi/abs/10.1029/2019RG000678.

[4] https://link.springer.com/article/10.1007/BF00143708.

[5] https://www.epa.gov/ghgemissions/global-greenhouse-gas-emissions-data.

[6] https://www.nationalgeographic.org/.

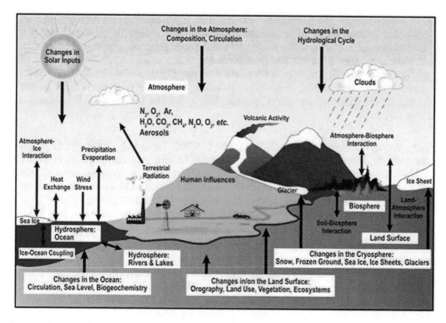

Fig. 1 Earth's climate, and a view of the inner workings of a climate model[7]

seriously contests that. Those who might wish that warming were caused by something other than humanity's use of fossil fuels have never proven another possible cause of that warming: If not fossil fuels, then what? No response to that question can be authoritatively answered and proven.

Figure 1 shows how a climate model works to illustrate major factors that determine the planet's climate. A climate model is an interactive and interconnected collection of mathematical relationships that characterize what goes on along each arrow in the figure. These models have many features in common with those used in generating daily weather projections, but many effects that can be assumed constant when looking forward only a few days – such as the relationship of the planet to the sun – vary substantially in the longer term.

In addition, a climate projection must consider accumulating greenhouse gases in the atmosphere that are "forcing" a change in that relationship.

Why are models necessary to explore whether or not human activity is contributing to the observed warming? The answer is simple: We have only one planet, and no separate "control planet" on which scientists can examine *everything except* human-caused forcing driven by fossil fuel consumption on Earth since the start of the industrial revolution.

---

[7] http://mpe.dimacs.rutgers.edu/images/climate-system-components/

One major source of confidence that much of the warming results from human emissions involves use of a standard procedure in scientific studies: counterfactual experimentation. Through that common process, climate models can provide estimates of what the global temperature would have been *with and without* the actual human activities and emissions. Reviewing results from coordinated runs of as many models as possible for these two conditions is the sound alternative to searching in vain for a control planet which of course does not exist.

The World Climate Research Program (WCRP) has, for some time, been conducting just that experiment. Its Coupled Model Intercomparison Project's (CMIP) fundamental mission is "to better understand past, present, and future climate changes arising from natural, unforced variability or in *response to changes in radiative forcing (from the emission of heat-trapping gases)* in a multi-model context." *(emphasis added)* This counterfactual exercise has been conducted simultaneously[8] by as many as 36 different modeling groups from around the world.

Why use so many models? Because each model on its own replicates the workings of such a complex climate system, as illustrated in Fig. 1. Each model attempts to characterize and quantify as many connections across as many climate variables as possible, thereby mimicking how the planet's climate actually works. Ensembles of model studies capture the implications of different methods used by the analysis teams, thereby exploiting the power of that diversity. In addition, ensembles allow researchers to compare models, analyze results statistically, and explore upper and lower extremes of potential climate change. Reviewed all together, they help scientists better understand what we do and do not know.

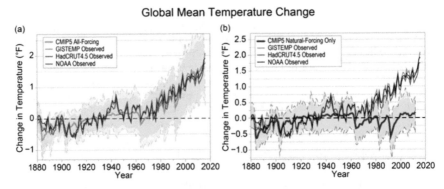

Fig. 2 Comparison[9] of observed global mean temperature anomalies from three observational datasets to CMIP5 climate model historical experiments using: **a** natural and anthropogenic and forcings combined, or **b** natural forcings only

[8] https://www.wcrp-climate.org/wgcm-cmip.

[9] https://science2017.globalchange.gov/chapter/3/#:~:text=Confidence%2520in%2520attribution%2520findings%2520of%2520anthropogenic%2520influence%2520is,confidence%2520for%2520severe%2520convective%2520storms%2520or%2520extratropical%2520storms.

Figure 2 shows some results from a model comparison that precisely investigates what causes observed warming.[10] It shows estimated series of actual temperature aggregates from NASA, NOAA, and UK Hadley Center measurement programs and compares them with the CMIP results.

Each of the 36 participating models in the CMIP exercise produced results for a "control earth" (natural forcings only) and also for the actual earth (accounting for all natural *and* human-caused forcings) from 1880 until today. In that way, each modeling group contributed to ensemble portraits of the two distinct forcing scenarios that can be compared with the actual historical record.

Panel (a), for example, displays the "all forcing" ensemble results for 36 models over time with its jagged orange shading. Panel (b), in contrast, displays the "natural forcings only" ensemble results with its blue shading.

Panel (a) shows that the "all forcings" ensemble results range above and below the actual global average temperature estimates. Taken as a whole, however, the ensemble average tracks the historical record very well year after year.

Panel (b) shows the range of ensemble results based only on natural forcings. Note that the counterfactual no-human-emissions temperature separates from the actual warming trajectories starting around 1980. Also, starting with a big dip in 1965, this ensemble average suggests that absent human influence, Earth actually would have cooled for a period up to around 2000. Our observations of what actually happened show that was not the case.

The bottom line is that scientists cannot explain the historical record without including human influences. So with very high confidence, the evidence convinces the science community both that the planet is warming and that human activity is largely to blame for that increased warming, particularly over the past seven decades or so

Afterword

Solid science stands behind the attribution of rising global temperatures to the greenhouse gas emissions resulting from human activities. But when that evidence is accepted, another critical question arises. What about local events associated with a changing climate: the forest fire that burned houses, the flood that swept away a town? Can specific climate events be attributed to the release of heat-trapping greenhouse gases, perhaps even providing a legal basis for holding emitters liable for damages? This is a question currently being raised in U.S. and international courts.

For damaging extreme events that have been associated with climate change, it is challenging to establish a direct causal connection to human

---

[10] https://yaleclimateconnections.org/2020/09/evidence-shows-the-planet-warming-on-average-at-an-increasing-rate/.

influence. In many cases, however, it can be shown that the likelihood of certain events, or categories of event, was enhanced by human activities. A useful analogy is a home run champion later found to have been taking steroids, known to increase batting prowess. Can it be shown that a home run in the third inning of a game against the Red Sox on July 18, 1987, was directly attributable to steroids? No, the batter who hit the home run was very good even without drugs. He might have hit that specific home run even if he were "clean." But the drugs increased the odds that he might knock one out of the park, especially in Fenway Park.

Essay 16. Extreme Events "Presage Worse to Come" in a Warming Climate

Prologue

At the time we drafted this essay, we were worried that explaining climate change risk solely in terms of temperature measurements and scientific studies would not be enough for some people. They have to "see it out of their own kitchen windows" to accept that the problem is worthy of their attention. Even if forced to concede that something unusual was happening to Earth's climate, they might argue that the effects are small and humans are resilient. So what if the daffodils in the backyard are showing up two weeks earlier? So what if the small brook is now flooding the back garden from time to time?

A growing number of damaging events in the U.S.—and globally—are providing plenty to see. Effects are visible on the TV, if not through the kitchen window. Costs of these events are no longer small. In this essay, we emphasize that extreme events are just the beginning of a very unfriendly future. As temperatures continue to rise, climate-related damages are going to mount. Nations cannot, on human time scales, expect a return to the climate that existed in the 1800s, before rapid industrialization. On the time span of a single human life, the detected increase in the frequency and intensity of observed extreme events is irreversible. We wanted to hammer home the point that the changes we are now seeing in the properties of extreme events reveal a stark vision of our choices for the future: abate, adapt, or suffer.

This essay first appeared as https://yaleclimateconnections.org/2020/10/extreme-events-presage-worse-to-come-in-a-warming-climate/.

G. Yohe et al., *Responding to the Climate Threat*, https://doi.org/10.1007/978-3-030-96372-9_16

Extreme events "presage worse to come" in a warming climate

Richard Richels, Henry Jacoby, and Gary Yohe

October 2, 2020

Mark Twain said a century ago that "fiction is obliged to stick to possibilities; truth isn't." Today we might just say, "You can't make this stuff up." However stated, the sentiment is painfully appropriate when it comes to extreme climate events. They are important not only for the damages they are inflicting here and now, but also for what they bode for the future. Many of the events we are currently experiencing would have been unimaginable in the latter half of the 20th century.

We would not have to turn the clock back very far to a time when few would have believed that a record-shattering heatwave in the western U.S. could raise Los Angeles temperatures above 120°F, or that areas of Siberia could experience temperatures of over 100°F. Nor would many have anticipated the wildfires, unrivaled in their intensity and destruction, in western U.S. states and Australia, or the Atlantic hurricanes of hitherto unexperienced ferocity, wreaking destruction of remarkable dimension, costs, and unpredictability. Even more surprising was the line of severe thunderstorms racing across the central U.S. with widespread winds of more than 100 mph, destroying crops and damaging structures. Pandemic concerns are heightened by warming, and quarantining compounds the challenges.

These are just a handful of the extreme events over just the past year, each augmented by a warming climate, causing pain and death and substantial economic loss. Unfortunately, the list could easily be expanded with other examples of severe storms, such as floods and droughts.

The list is truly mind-numbing, and understanding what it means is greatly aided by the meticulous record keeping of federal agencies such as the National Oceanic and Atmospheric Administration (NOAA). Its website has kept track of extreme climate events exceeding billions of dollars. Figure 1 shows the trend of the past four decades in the U.S., where the overall damage costs reached or exceeded $1 billion (all adjusted to equivalent 2020 dollars). Definitely cause for concern.

As if the extreme weather events to date weren't enough, so far in 2020 the globe has experienced another type of extreme event related to climate change which until recently had escaped the public's radar screen, one with a unique ability to cause widespread planetary harm in a matter of weeks. It now appears[1] that the likelihood and cost of a pandemic, such as COVID-19, is increased by climate change. The natural barriers that separate humans from the wild have been weakened by a loss of natural forests, one effect of a

---

[1] https://thehill.com/opinion/energy-environment/499604-we-cannot-ignore-the-links-between-covid-19-and-the-warming-planet.

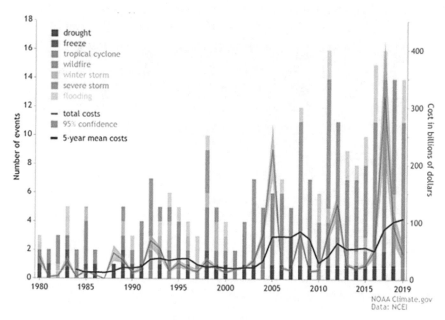

Fig. 1 U.S. billion-dollar disasters by type, 1980–2019[2]

warming climate, increasing the possibility that the contagion can jump from one species to another.[3]

Once humans become infected, containment is made more difficult by other extreme events. Indeed, the American Red Cross now calls for hurricane preparedness to consider the needs[4] for quarantining, social distancing, and other measures aimed at preventing a new COVID-19 hotspot. Preparations for other extreme climate events will require similar rethinking.

Why are we seeing all these extreme events? In two previous essays as part of this series [see here[5] and here[6]], we summarized scientific evidence on warming and on its being caused primarily by human emissions. Of course, it would not be right to blame every individual weather event on global warming. But extreme hot days will become more likely, and severe cold snaps less so, and

[2] https://www.climate.gov/news-features/blogs/beyond-data/2010-2019-landmark-decade-us-billion-dollar-weather-and-climate.

[3] https://www.nature.com/articles/d41586-020-02341-1#:~:text=As%2520humans%2520diminish%2520biodiversity%2520by,pandemics%2520such%2520as%2520COVID%252D19.&text=Jones%2520is%2520one%2520of%2520a,use%2520and%2520emerging%2520infectious%2520diseases.

[4] https://www.redcross.org/get-help/how-to-prepare-for-emergencies/types-of-emergencies/coronavirus-safety/preparing-for-disaster-during-covid-19.html.

[5] https://yaleclimateconnections.org/2020/09/evidence-shows-the-planet-warming-on-average-at-an-increasing-rate/.

[6] https://yaleclimateconnections.org/2020/09/the-evidence-is-compelling-on-human-activity-as-the-principal-cause-of-global-warming/.

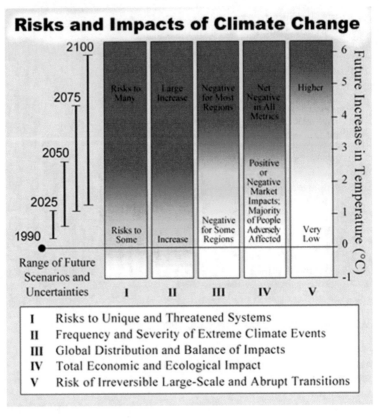

Fig. 2 Reasons for concern about climate change[7]

with the shift to warmer temperature patterns come the changes in flood, drought, fire, loss of glacial ice, and more.

Today's extreme weather events 'a troubling harbinger'

While the resulting climate extremes experienced so far are cause enough for profound concern, scientists have been warning that they are a canary in the coal mine, a troubling harbinger of worse to come. Figure 2 depicts the first of a continuing series of such diagrams entitled "Reasons for Concern" (RFC) in climate studies. It shows extreme events (RFC-II) as early precursors of even more profound threats to the habitability of the planet. These include irreversible large-scale and abrupt transitions (RFC-V) that occur when positive feedbacks inadvertently lead to a tipping point – such as the loss of the Amazon rainforest or rapid melting of polar ice sheets – potentially with catastrophic consequences.

---

[7] https://www.ipcc.ch/site/assets/uploads/2018/05/SYR_TAR_full_report.pdf.

The processes leading to potential tipping points[8] were set in motion by the burning of fossil fuels and deforestation in the industrial revolution. The warming that has already occurred is leading to the thawing of permafrost, opening the way for the release of huge amounts of additional greenhouse gas into the atmosphere, both methane and CO_2. Continued warming also threatens additional harm to forests and their ability to sequester CO_2 and accelerates the melting of Arctic ice, which reflects the incoming solar heating.

These and other positive feedbacks in the climate system serve to further amplify warming and sea-level rise. Scientists warn that Earth's large-scale ocean circulation, which is a major planetary heat-dispersion system, may be further compromised, both in the North Atlantic and beyond. Such warnings are ignored at great peril.

If all of this is disturbing, it should be ... because it is. The extreme events that we are witnessing are not only worrying in their own right, but presage worse to come. If one asks why we still procrastinate to do something about it, good question. In the case of climate change science – as with all science – there will always be some lingering uncertainty. If not about the likelihood, then about the consequences. Or if not about the consequences than about who is at fault and should bear the burden. But keep in mind that uncertainty works both ways, both in overestimating and in underestimating potential risks. In the case of climate change, it appears that scientists have tended to under-, rather than over-, estimate the pace of change. Either way, continuing to kick the can down the road should not be an acceptable strategy. Surely the time for procrastination is over.

Action must initially involve three parties: the scientists whose job is to enlighten and advise; the electorate whose job is to decide how risk averse they choose to be; and their political leaders whose job is to reflect the will of the people and act accordingly. The canary in the mine mouth is on life support. Its fate is in all of our hands. We must all do our jobs.

Afterword

Just a few months after this essay appeared, the National Oceanographic and Atmospheric Administration (NOAA) released an updated summary of major U.S. events whose likelihood and/or scale had been increased by rising temperatures. The summary focused on individual events that had caused at least $1 billion in damage. In 2021, the upward trend in these damaging events continued, as Hurricane Ida laid down a swath of destruction from the Gulf Coast to New York and New Jersey. 2021 was also another year of record fires ravaging Western states. The first half of 2022 was marked by extreme

---

[8] https://www.nature.com/articles/d41586-019-03595-0.

drought across the southwestern states and California, unprecedented wild-fires in Alaska, and record heat in the southern half of the country, from the Rockies well past the Mississippi River.

These are only the well-identified expensive losses. Missing in the NOAA summaries are the costs of many small events. Also missing are the mounting expenditures in small communities to increase resilience to climate risk.

The federal government can help with both mitigation and adaptation efforts. For example, under the Biden-Harris Administration, the U.S. is on a path toward major improvements in the nation's infrastructure. Global warming and the associated sea level rise will call for new designs of roads, bridges, and water systems to prepare for the flooding arising from more severe and frequent storms.

Essay 17. Multiple Extreme Climate Events Can Combine to Produce Catastrophic Damages

Prologue

In this next essay, we took another stab at conveying an understanding of the complexity of changes brought about by a warming of the globe. For ease of understanding, portraits of the environmental, economic, and social implications of a changing climate tend to focus on one potential effect at a time—rising sea level on a particular section of coastline, hotter summers in Chicago, dying coral reefs, or methane releases from the tundra. The stark reality is that these events are occurring simultaneously.

Some of the most damaging consequences arise because different aspects of a changing climate collide with one another at the same place and at the same time, producing damage far greater than what would be expected from each influence itself. Drought, extreme heat, and devastating wildfires in the western U.S. were prominent in the news in 2020, and they are prime examples of this phenomenon of a "nexus of climate risk." We wanted to explain how certain features of a warming climate combined to produce this nexus.

This essay first appeared as https://yaleclimateconnections.org/2020/10/vigorous-action-needed-and-soon-on-climate-change/.

G. Yohe et al., *Responding to the Climate Threat*, https://doi.org/10.1007/978-3-030-96372-9_17

Multiple extreme climate events can combine to produce catastrophic damages

Gary Yohe, Henry Jacoby, and Richard Richels

October 9, 2020

Wildfires in California, Oregon, and Washington are this year's poster children for extreme natural disasters. Hardly a day passed in August and September without disturbing pictures of heart-wrenching damages and loss of life.[1] Even worse, this summer's hurricanes became major flooding events[2] as the storms themselves stalled overpopulated areas along the Gulf coast.

That does not mean, of course, that all see climate change as playing a significant role in determining the strength, frequency, or behavior of either of these climate risks.

What it *does* mean is that the scientific community must explain more clearly why the recent spate of extraordinary natural disasters can be understood only with reference both to impacts of climate change as we have come to know them,[3] and now something more complex: concurrent impacts amplifying themselves in real time.

Figure 2 of our September 18th essay[4] in this series showed how global warming can push aspects of the environment toward greater extremes and higher damages. More specifically, it teaches us how trends that increase damages can, over time, make high-damage futures more likely while reducing the chances of more benign possibilities.

Recent events have taught us more than that, though. They have demonstrated a troubling propensity for several climate change impacts to show up at the same place at the same time, feeding on each other, combining forces and leading to still greater extremes. To be clear, they do not necessarily arrive at the same time and/or leave at the same time; but they do spend a significant amount of time together compounding their extreme impacts on a specific location.

The California fires are a perfect example of this phenomenon.[5] Only three of the state's largest 20 fires (in terms of acres burned) had burned prior to 2000, but nine of the biggest 10 have occurred since 2012. That is, extremes are becoming more likely. And they are growing larger too. In 2017, 9,270 fires burned a record 1.5 million acres. The Mendocino Complex fire the next year became the "largest wildfire in California history." And soon came 2020.

---

[1] https://www.theatlantic.com/photo/2020/09/photos-wildfires-rage-across-american-west/616219/.

[2] https://en.wikipedia.org/wiki/2020_Atlantic_hurricane_season.

[3] https://www.sciline.org/quick-facts.

[4] https://yaleclimateconnections.org/2020/09/evidence-shows-the-planet-warming-on-average-at-an-increasing-rate/.

[5] https://www.fire.ca.gov/incidents/2020/.

A new largest fire in California history, the Complex fire, got started in August 2020. Soon came the 3rd, 4th, 5th, and 6th largest in history. By October 3, these five conflagrations had combined with nearly 8,000 other more "ordinary" fires to kill 31 people and burn more than four-million acres, and, on that day, all five of those fires were still burning.

Why is this happening? Wildfire is a natural part of the forest environment. But by the early 1950s, these fires were causing sufficient damage with sufficient frequency to provoke efforts to reduce what was seen as the main cause, human behavior. "Only you can prevent forest fires" was the mantra of the times.

Only a few decades later, however, changes in the climate had begun to contribute[6] to increased fire risk. More intense droughts played a role in some years, as did extra strong heat waves. Also, milder winter temperatures were fostering the expansion of a major forest pest, the Pine Bark Beetle, which was killing large areas of forest and thereby further increasing the supply of ready fuel.

Of course, part of the increased fire risk is still the result of human actions.[7] Damage to life and property has increased markedly as more people have moved into vulnerable forested areas, and more people in the woods means more inadvertent blazes. Changes in forest management contributed, too, because fire suppression policies on federal land reduced the brush-clearing value of deliberately set control blazes (sometimes known as "good fires").

But these non-climate causes of increased fire danger have not increased so much over the decades to account for the devastation of the last few years. There is more to explain, and it comes in understanding how, in responding to rising global temperatures, nature can produce "2-fer" or even "3-fer" combinations of influences on local environmental conditions.

The western U.S. is, unfortunately, a clear example of this effect. Many of the fires were caused by literally thousands of dry lightning strikes.[8] These strikes aren't the result solely of climate change, but it is clear that they fed into a witches' brew of conditions that are all linked to global warming:

1. the lightning strikes and other points of ignition in the midst of a record drought;
2. record heat for days on end in July and August;
3. infestations of bark beetles producing large stands of dead trees; and
4. decades of gradual warming extending the western fire season by some 75 days.

Taken together, these contemporaneous impacts make it clear that the issue is not just what sparks the fires. The larger problem is the context in which

---

[6] https://www.fire.ca.gov/incidents/2020/.

[7] https://www.nytimes.com/article/why-does-california-have-wildfires.html.

[8] https://www.latimes.com/california/story/2020-08-19/destructive-bay-area-fires-fueled-by-rare-mix-of-intense-dry-lightning-and-extreme-heat.

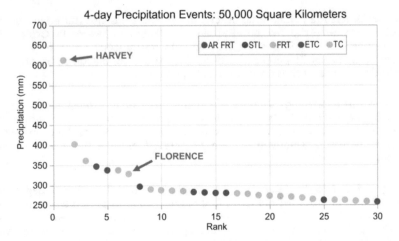

Fig. 1 Heavy precipitation events in the contiguous United States. Averaged over four days and over 50,000 km² through 2019. Hurricane Harvey produced the most precipitation by far, while Florence ranked seventh (and 2nd among tropical cyclones). Other categories include fronts associated with an extratropical cyclone (FRT), extratropical cyclones not color coded with fronts (ETC), fronts with an associated atmospheric river (AR FRT), and subtropical lows (STL). *Source* North Carolina Institute for Climate Studies[10])

they start, and how quickly they spread once started, especially when several intensifying influences are also present.

A similar story can also be told about damages from tropical storms. As shown in Fig. 1, hurricanes Harvey and Florence dropped historic amounts[9] of rain after making landfall and then stalling over Houston and New Orleans, respectively. This summer, hurricanes Laura and Beta followed suit, dumped extreme rainfall totals, and caused substantial damage from storm surge.

Here is another example of a climate change induced compound effect – a "3-fer":

1. near record-warm ocean temperatures allowed many tropical depressions and non-tropical low-pressure systems to develop into dangerous hurricanes;
2. a decrease in the summer temperature difference between the Arctic and the tropics that is strongly suspected[11] to have weakened atmospheric steering currents and created more slowly moving storms.
3. sea-level rise, one of the most obvious results of decades of rising temperatures, compounded risks posed by storm surge.

---

[9] https://www.washingtonpost.com/news/wonk/wp/2017/08/29/houston-is-experiencing-its-third-500-year-flood-in-3-years-how-is-that-possible/.

[10] https://ncics.org/cics-news/hurricanes-harvey-and-florence-historical-context/.

[11] https://theconversation.com/what-makes-huricanes-stall-and-why-is-that-so-hard-to-forecast-146804.

The expanding consequences of compound fire and flood events are also getting harder to control and survive. For example, many of the worst fires and hurricanes have exploded so quickly and spread so erratically that human evacuations have become "moment's notice" emergencies. Just as with residents of the southeastern and Gulf coasts, residents from California and Oregon must retreat from harm's way as quickly as possible, and hope that conditions will soon change back to something more benign.

Over time, the weather eventually becomes more favorable. Unfortunately, the climate is not going to change back to what used to be, certainly not on a human time scale. So, when they have a chance, perhaps vulnerable residents should just try to move as far from harm's way as possible.

That might be a good idea for the short-run, but lest we forget: None of us can move to a different planet.

Afterword

Because of word limits, there were other instances of this interaction among climate change impacts that we could not cover. For example, in areas where drier conditions are reducing crop output, rain is frequently concentrated in more intense downpours. The deluges not only cause further damage to crops; they also wash out farm-to-market roads.[12] These effects can combine to do more damage to the farm economy than would be estimated as the sum of the damages from each individual impact.

The oceans are another prime example of intersecting and overlapping impacts. First, they are soaking up a large fraction of human-caused warming, and the resulting higher ocean temperatures affect sea life. This effect is seen most dramatically in the bleaching of corals, but also manifests itself in the migration of species. Second, the oceans are also absorbing a significant fraction of the carbon dioxide produced by fossil fuel burning. This absorption results in an increase in ocean acidity (a decline in the pH of ocean surface waters). Ocean acidification can have potentially dramatic effects on sea life, including on microorganisms like phytoplankton[13] that are at the base of the marine food web. The interaction of these two effects, heating and acidity, could cause even greater damage than the sum of expected damages estimated in isolation of interaction.

---

[12] https://www.wider.unu.edu/publication/economic-impact-climate-change-road-infrastructure-sub-saharan-africa-countries.

[13] https://news.mit.edu/2015/ocean-acidification-phytoplankton-0720.

Essay 18. Vigorous Action Needed, and Soon, on Climate Change

Prologue

Hoping that our earlier articles would draw attention to the threat of human-caused climate change, we turned to address the question of what can be done about it and the urgent need for an effective response. As noted earlier, climate change deniers frequently argue that it is too expensive to reduce the greenhouse gas emissions that are the root cause of the threat. An addendum to this argument is that, given the scientific uncertainties, the nation should delay action until we have more complete understanding of the climate threat. And, of course, the U.S. should not get out ahead of China and take unilateral actions!

Our task in this essay was to emphasize the high cost of delaying concerted efforts to reduce CO_2 emissions. We also wanted to point out that the U.S. was responsible for the largest share of the historical emissions of greenhouse gases. In our opinion, our country must therefore drive the process of finding innovative solutions to this problem. We cannot simply be a greenhouse gas "free-rider," tagging along on the coat tails of other nations.

This essay first appeared as https://yaleclimateconnections.org/2020/10/vigorous-action-needed-and-soon-on-climate-change/.

Vigorous action needed, and soon, on climate change

Henry Jacoby, Gary Yohe, and Richard Richels

October 16, 2020

Policy and politics

Our essays in this series have presented compelling scientific evidence about the warming of the planet, reviewed the evidence that human activity is its principal cause, and discussed the resulting economic and environmental damages.

Now comes the question of what we are going to do about it. The options are clear:

- Nations can work toward eliminating greenhouse gas emissions and reducing the scale of future warming.
- Governments and private actors can, and will, invest in measures to protect home and livelihood from effects of changes that cannot be prevented.
- Or human societies and natural ecosystems will suffer the severe harms of inaction.

The more they (really we) do now and in the near future, the smaller will be the residual damages imposed on ourselves, our children, and our grandchildren. The choice is ours.

The suffering is already here, of course. In some places, it is almost impossible to bear despite growing investments in adaptation.[1] So what is missing? A commitment to emissions reductions appropriate to the *special nature* of the climate change threat. Fortunately, with a smart choice of policy measures, the emissions control challenge can still be met at a tolerable economic cost.

What makes threats posed by climate change 'special'?

And what is the special nature of the threat? Climate change is unlike other environmental insults, such as polluted urban air. Risks associated with most pollutants depend on current emissions whose corrosive properties and damages are generally reversible. Not so for our influence on the climate. Human emissions contribute to stocks of heat-trapping gases in the atmosphere, some with very long residence times there. Carbon dioxide (CO_2) is the most important of these gases, and natural processes cannot reduce its concentration[2] resulting from human emissions except over centuries to millennia.

---

[1] https://www.wri.org/blog/2015/04/costs-climate-adaptation-explained-4-infographics.
[2] https://www.pnas.org/content/pnas/106/6/1704.full.pdf.

Over the past several thousand years, Earth's temperature has varied within a narrow range of less than 1 °C, about 1.6 °F[3] – stable conditions under which human society has thrived.

Spurred by human emissions of greenhouse gases, however, the planet's temperature has increased by 1.1 °C in a flash, a bit over one century. The current economic and environmental damages are the result of just the ongoing warming of the planet so far; that warming is spurred by greenhouse emissions since the industrial revolution, and it cannot be reversed on human time-scales. And each year of continued emissions commits Earth to ever-higher temperatures.

Avoiding 'potentially calamitous' warming

The need for an urgent response is made clear by even a cursory look at what is a stake if this human influence on the climate is allowed to continue unabated. Figure 1 displays the latest version of the "burning embers" diagram,[4] picturing reasons for concern with climate warming, as introduced in our October 2nd essay. As displayed in the first four columns, warming of 2 °C would be more damaging than the current 1.1 °C increase. Further increases would be worse, and the still higher increases potentially calamitous.

More troubling, even the damages we know a good deal about, and can expect at these higher temperatures, do not tell the whole story. There are risks, still poorly understood, of so-called tipping points, suggested by the right-most column in the figure, in which case the Earth might undergo a radical environmental shift. Examples[5] include an unstoppable melting of Greenland and the Antarctic ice sheets that will yield many meters of sea-level rise, or an acceleration in the outgassing of CO_2 and methane (*even less well understood* for an especially potent greenhouse gas*) from tundra ecosystems that would rapidly increase the pace of warming.

Needed soon – 'tenacious, long-term effort' to control greenhouse gas emissions

Continued greenhouse emissions will push these risks higher and higher, and managing them demands a prompt, vigorous emissions control effort. In framing a response there is no simple temperature goal or emissions target that will determine success or failure of the effort. A focus on such a mistaken, do-or-die, achievement, if some particular target appears unlikely to be met, would risk causing despair, and a shift elsewhere, of public energies.[6] Managing the climate threat will require a tenacious, long-term effort to limit greenhouse emissions whatever the level of achievement at any time along the way.

---

[3] http://www.realclimate.org/index.php/archives/2013/09/paleoclimate-the-end-of-the-holocene.

[4] https://www.nature.com/articles/nclimate3179.

[5] https://www.nature.com/articles/d41586-019-03595-0.

[6] https://www.newyorker.com/culture/cultural-comment/what-if-we-stopped-pretending.

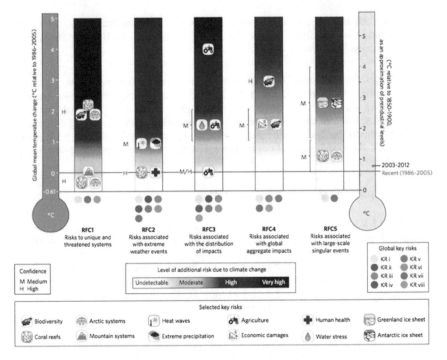

Fig. 1 Enhanced burning embers diagram, providing a global perspective on climate-related risks[7]

Of course, as economists are wont to say, there is no free lunch. There will be costs to the economy in the transition away from fossil fuels and in the cutting of other greenhouse emissions. Fortunately, studies of long-term climate policy[8] routinely find that, with effective international cooperation, deep reductions in global emissions could be achieved over time with only a few percent loss in economic activity. And these estimates don't account for the costs avoided by the lowering of future climate change.

Moreover, there is continuing improvement in technology and policy design. For example, the costs of low-carbon technologies like wind and solar power[9] continue to fall, and economically efficient emissions pricing initiatives are being ever more widely applied[10] to wean economies off fossil fuels. One indication of growing confidence that costs are manageable is that the European Union[11] and China[12] talk of reducing their emissions to zero by around mid-century.

---

[7] https://www.nature.com/articles/nclimate3179

[8] https://www.ipcc.ch/site/assets/uploads/2018/02/ipcc_wg3_ar5_full.pdf.

[9] https://blogs.imf.org/2019/04/26/falling-costs-make-wind-solar-more-affordable/.

[10] https://carbonpricingdashboard.worldbank.org/.

[11] https://www.wsj.com/articles/eu-to-cut-greenhouse-gas-emissions-to-zero-by-2050-11576203017.

[12] https://www.nytimes.com/2020/09/23/world/asia/china-climate-change.html.

Whatever the overall economic costs of cutting warming emissions may turn out to be, the greatest burdens will fall on a narrow set of businesses, fuel producing regions, and employment groups. In a number of U.S. states and regions, the prerequisites for aggressive emissions reduction likely include programs to ameliorate these short-term impacts, perhaps including extended unemployment support and training for workers whose jobs disappear permanently.

As the second largest emitter of greenhouse gases and world power, the U.S. and its response to the climate challenge are crucial. Evaluation of U.S. performance to date depends on where you look. On the one hand, many individuals, non-profit organizations and business firms are actively pursuing emissions-reducing investments and changes in operating practices. Also, states and cities[13] have adopted aggressive emission targets and are implementing policies and programs to meet them.

On the other hand, the achievements of these earnest efforts are limited by a lack of supportive, coordinated emissions policies from the U.S. federal government. Still more troubling, one side-effect of this domestic inaction is a failure of national leadership within the global climate effort, a topic for our next essay.

Our message, then, is clear. With intelligent actions, strenuous emissions control policies will impose manageable costs on the overall economy, though the energy transition will be more painful for some business sectors and regions. Not taking prompt action would, however, be very expensive. Further delay raises the risk of ever-increasing, perhaps catastrophic, environmental and economic damages.

Editor's note: The authors edited this piece on October 19 to add and emphasize the words "less well understood" as it applies to methane. They did so in response to an email from Scripps Institution of Oceanography climate scientist Jeffrey Severinghaus concerning a recent Science article[14] of which he was a coauthor. He and his coauthors advise against seeing "methane, from tundra, [a]s a significant source of our future climate forcing." The authors of the essay above based their original comment on language in the IPCC's Fifth Assessment Report (p. 531[15]). That language reads as follows: "Should a sizable fraction of this Arctic frozen carbon be released as methane and CO_2, it would increase atmospheric concentrations, which would lead to higher atmospheric temperatures. That in turn would cause yet more methane and CO_2 to be released, creating a positive feedback, which would further amplify global warming."

The authors express their thanks to Professor Severinghaus, saying his paper "adds to the knowledge base by providing more quality data to the issue and an argument that outgassing contributions might not be so important as was previously thought. Assessments of confidence in science findings like 'this

---

[13] https://www.c2es.org/content/state-climate-policy/.

[14] https://science.sciencemag.org/content/367/6480/907.

[15] https://www.ipcc.ch/site/assets/uploads/201802/WG1AR5_all_final.pdf.

might happen' depend on the quality of the evidence[16] *(data) and agreement about underlying processes. This new science therefore reduces confidence in the future occurrence of what would be a high-consequence event by lowering assessments about its likelihood; but it does not negate its appropriate inclusion in elaborating a reason for concern. Rather, it identifies an area for future research and increases the importance of working to better understand what is going on and what might happen downstream."*

Afterword

In the 2020 election, which motivated this series of articles, action on climate change was only one of many areas of controversy between the opponents. Candidate Biden and the Democratic party pledged action on climate change while Candidate Trump and his party did not. Climate did not turn out to be a major issue in the 2020 campaign. Unfortunately, the Biden victory failed to result in immediate and effective action on climate. Mounting a federal response proved very difficult, and the urgency of the essay's message has only increased with time.

Though we did not plan it, this essay also provided a lesson in how the scientific process works. New information is constantly being incorporated in successive assessments of issues featured in this essay—not just new understanding of the magnitude of the threat and the cost of fighting it, but also new insights into the underlying science. The inclusion of a footnote after the essay was published is an example of the dynamic nature of the scientific process. Citing improved data and new analysis, a reader recommended that we should revise the confidence attributed to estimates of the outgassing of methane from the Arctic. The issue is important and the science is still in development. We used the reader's comment as a teachable moment—we improved our discussion of methane release and explained "how science works" to readers.

---

[16] https://archive/ipcc.ch/pdf/supporting-material/unertainty-guidnce-note.pdf.

Essay 19. Rejoining the Fight Against Climate Change is in the U.S. National Interest

Prologue

While writing this series of weekly essays, we frequently heard the mantra of "America First!" So did millions of other Americans. It was important to point out the danger in this perspective if it blinds the U.S. to areas where our interests depend upon the actions of other nations. The U.S. is a world economic power. Based on our historical cumulative emissions, the U.S. is the largest contributor to the current atmospheric load of heat-trapping greenhouse gases. It remains the second biggest emitter today (after China). The U.S. could either be a supporter (perhaps even leader) of international efforts to limit emissions, or it could be an overwhelming drag on these efforts.

We also felt it was essential to emphasize that significant cuts to U.S. emissions will not solve the global problem of avoiding dangerous anthropogenic interference with our climate system. The U.S. is a big emitter, but not that big. We are but one part of the global problem. However, by its domestic actions, the U.S. is capable of sending loud messages to the rest of the world. In our view, these messages should be crystal clear: that the U.S. is not free-riding on the efforts of others, but is instead helping to correct the global inequities that hamper the needed universal commitment to action.

This essay appeared first as https://yaleclimateconnections.org/2020/10/rejoining-the-global-fight-against-climate-change-in-the-u-s-national-interest/.

© The Author(s), under exclusive license to Springer Nature Switzerland AG 2023
G. Yohe et al., *Responding to the Climate Threat*,
https://doi.org/10.1007/978-3-030-96372-9_19

Rejoining the fight against climate change is in the U.S. national interest

Richard Richels, Henry Jacoby, and Gary Yohe

October 23, 2020

When the Trump Administration gave notice that the U.S. would drop out of the Paris Climate Agreement, it said it was doing so because it was a bad deal for the country. This view is wrongheaded. The science is unequivocal. Global warming is real, human induced and, if unabated, it poses catastrophic risks to all inhabitants of the planet.

Limiting this threat requires a global effort, but that effort is currently hampered by the absence of the U.S. as a major world power and second largest source of greenhouse gases (and leader in terms of historical CO_2 emissions). It is in our national interest, as well as the right thing to do, not just to contribute to the global effort, but also to take an active role in its evolution.

To believe otherwise belies the nature of the threat and what it will take to manage it: The total of *global* emissions of greenhouse gases must be eliminated, and all the nations[1] of the world are contributing to it. Therefore, reducing the domestic threat will require universal action, and inaction by the U.S. encourages free riding by others. If we are not doing our part, why should they?

How much will continuing delay[2] cost? The National Oceanographic and Atmospheric Administration (NOAA) has kept careful track of extreme climate events over the past four decades. Damages thus far come to trillions of dollars; and they are growing. But these include only part of the picture, what is referred to as "market" damages – expenditures that show up in national income accounts as a result of hurricanes, heatwaves, wildfires, severe storms, floods, and droughts.

Even more worrisome, though, are "non-market" impacts such as mortality and morbidity where assigning a dollar value[3] is much more difficult and contentious. Also included in this category are losses in environmental quality and biodiversity and loss of ecosystems upon which our future prosperity depends.

Taken all together, the risks are, indeed, frightening. And what we have experienced so far are likely only a forewarning of worse to come as ice sheets melt, forests weaken and decline, permafrost thaws releasing carbon, and sea level rises. By any measure, the stakes are extraordinarily high.

So, what should our political leaders do to meet their obligations to protect the health, safety, and welfare of the country? No doubt much effort will

---

[1] https://www.ipcc.ch/sr15/chapter/spm/.

[2] https://www.ncdc.noaa.gov/billions/.

[3] https://www.oecd.org/env/cc/2483779.pdf.

be required to re-establish international trust and leadership so recklessly discarded over the past nearly four years, but three steps are necessary if we are to rejoin the global effort to meet the climate challenge.

Step 1 will require immediately signaling other nations that we are ready to resume our role in the global climate effort by rejoining the Paris Climate Agreement. Adopting emission reduction goals and implementing actions to meet them, comparable to those already under way among other wealthy participants, will be an essential part of this step. These commitments need to look beyond the initial 2030 focus of the Paris agreement to establish firm policies for longer-term emissions reduction. Also, thought needs to be given to preventing future administrations from arbitrarily and capriciously reversing course.

Anyone concerned that the costs of achieving a non-carbon economy are too high need only study the consequences of not doing so. The investment will be a bargain, which is not to imply that eliminating U.S. emissions will be costless. The transition will necessitate early retirement of existing power plants and equipment that, but for their contribution to global warming, would have years of remaining life-time. The costs of replacement will not be insignificant, but in the long term the reduction in damages will justify the investment.

The question is how to achieve such an outcome. The first round of industrialization was powered by fossil fuels, then an "inexpensive" and plentiful source of energy. Only later did the full costs become apparent. Future technology choices must reflect the true costs of energy, not only to the economy but also to public health and wellbeing. Recognizing costs – both market and non-market – that are currently excluded from the calculus will help to tilt technology choice decisions in the optimum direction.

Step 2 Fig. 1 shows that the U.S., with 4% of the world's population, is responsible for approximately 15% of global CO_2 emissions. China[4] and many other industrialized countries[5] have pledged dramatic emission reductions by mid-century. The intentions of many lesser developed countries, aspiring to a lifestyle akin to their wealthier neighbors, are less well defined. If these countries opt for development paths powered by fossil fuels, emission control efforts of industrialized countries will not stop the warming. Total global emissions could stay roughly the same; only their geographical origins would change.

Incentivizing developing countries to forgo carbon-polluting technologies in the end will be a heavy lift. Without a clear commitment to re-engage in the effort to halt global warming, and stick to it, the U.S. will be in a weak position to persuade other nations to reduce their emissions. In addition to moral and political support, developing countries will need technical and financial assistance.

Technology development and transfer, then, is one area where the U.S. can help developing nations choose a low-carbon development path, with the U.S.

---

[4] https://www.nytimes.com/2020/09/22/climate/china-emissions.html.

[5] https://unfccc.int/sites/default/files/resource/docs/2014/sbsta/eng/inf06.pdf.

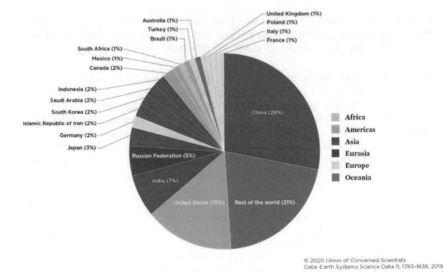

Fig. 1 National shares of global 2020 CO_2 emissions[6]

providing technical expertise and funding support. Moreover, active involvement in these areas will open up opportunities for U.S. participation in important emerging markets.

Step 3 will involve the massive and immediate collective action required to manage the impacts of climate threats, and the U.S. focus should not be limited to emissions abatement. Nations must also deal with the pain and suffering that inhabitants of the planet are currently experiencing and that will worsen given the additional warming already baked into the system. Active contribution to these adaptation efforts is also in the U.S. national interest.

Just as greenhouse gases respect no national boundaries, neither do their impacts. If warming continues unabated, all will be subjected to its increasing damages, with the impacts most harshly felt by those least able to deal with them. Without help to address the challenge, global political stability will be at risk, leading to millions of environmental refugees and stateless people[7] being put at great risk. Regional conflicts and humanitarian crises then could exhaust the resources of rich and poor countries alike. One can envision a domino effect spilling over vast areas of the planet.

The U.S. is at a fork in the road. We can continue to deny our responsibility for this problem, and in so doing risk a future fraught with suffering and pain. Or we can rejoin the international effort and help open the door to a brighter future. These efforts will involve not only putting our own house in order, but assisting others to do the same.

---

[6] https://www.ucsusa.org/resources/each-countrys-share-co2-emissions

[7] https://www.brookings.edu/wp-content/uploads/2019/07/Brookings_Blum_2019_climate.pdf.

Some may derisively dismiss such efforts as merely "humane." So be it. But rest assured they are also in our own self-interest.

Afterword

From its first day in office, the Biden-Harris Administration emphasized that U.S. interest lies in domestic action to reduce emissions, and in an active diplomatic role in the international emissions reduction efforts. On the domestic side, a 2030 emissions target was set. This target puts the U.S. in line with pledges made by other developed nations. Policy changes were proposed to meet the target.

The U.S. international role was strengthened by rejoining the Paris Agreement and appointing a cabinet-level climate envoy (John Kerry) to focus on the nation's role in global negotiations. In 2021, the president used meetings with global leaders to highlight the U.S. commitment to emissions reduction and to call for action by others. These meetings included a Leaders Summit on Climate Change (on Earth Day 2021) and the Major Economies Forum on Energy and Climate. The U.S. also increased its planned contribution to funds for aiding the emissions reduction and adaptation efforts of developing countries.

This marked a dramatic shift, from one administration to the next, in terms of how seriously climate change was viewed. But this shift was only the beginning. Decades of difficult work will be required to meet U.S. emissions targets. The nation's current commitments of technical and financial aid to developing countries fall well short of what is actually required. There is also justifiable international skepticism about U.S. willingness to "stay the course." In the case of the Kyoto Protocol and the Paris Agreement, the U.S. helped to negotiate a deal, and then failed to ratify it. With the potential for future shifts in the U.S. political balance, there is risk of once again losing sight of the national interest in fighting climate change. That national interest is clear—the U.S. loses if the world fails to address human-caused climate change.

Part V

Early Months of the Biden-Harris Administration

During the presidential campaign, candidate Biden declared his support for federal action on climate change. He also pledged that under his Administration, the U.S. would resume its appropriate role in global climate negotiations by rejoining the Paris Climate Agreement and updating the U.S. emissions pledge. Congressional legislation was proposed to meet these obligations after Biden's election to the Presidency, but it soon became clear that the Senate would be a formidable obstacle to progress on climate.

There were, of course, other federal actions that could still be accomplished, and in four essays we called attention to a few of them. The first three essays concern actions easily within the Biden-Harris Administration's grasp—offering guidance for a nearly completed federal study of U.S. climate risk (Essay 20), calling for attention to a time-limited opportunity to reverse certain Trump-era climate policies (Essay 21), and highlighting the strong public support for executive action on climate (Essay 22). The last essay in this quartet argues for the advantages of a carbon tax (Essay 23), which was not on the Administration's agenda in early 2021.

Another two essays concerned the need to enhance respect for climate science—an essential ingredient in order to grow public support for Biden's ambitious climate agenda. In one of these essays (Essay 24), we note President Biden's actions to reestablish respect for science, undermined in the Trump years, through Biden's appointments of highly qualified individuals to key government posts and through the elevation of the President's science advisor to cabinet rank. The other essay in this pair (Essay 25) takes on a much-hyped book which dismissed climate change as a serious policy concern because it claimed that the science was "unsettled." Not so, we argue, particularly given that the author of the book failed to examine the science from a risk perspective.

Policy concerns in this period continued to be stimulated by ever more damaging extreme weather events. We wrote two essays in response to these events. One explores connections between global warming and extreme weather (Essay 26). The other examines recent extreme events that are far outside of any recorded experience and explains why they can only be explained by taking climate change into account (Essay 27).

Essay 20. Early Next Step: Add Risk Management to the National Climate Assessment

Prologue

In 1990, the U.S. Congress established a research program on global change and required a periodic report that would assess current and long-term trends (mainly in the climate) and their effects on the U.S. The fifth of these National Climate Assessments (NCAs) is scheduled for release in 2023. Work on it had largely proceeded under the radar during the Trump years, but with the incoming Biden-Harris Administration, new life would be injected into the effort. We saw an opportunity to expand the scope of the report—making it more useful in public discussion of climate policy by extending its analysis into potential management responses. While previous NCAs provided detailed summaries of the risks of a changing climate for U.S. regions and sectors of the economy, including a general discussion of ways to reduce these risks, those earlier assessments always stopped short of analyzing specific public response measures.

This essay first appeared as https://yaleclimateconnections.org/2021/01/commentary-early-next-step-add-risk-management-to-national-climate-assessment/.

Early next step: Add risk management to National Climate Assessment

Gary Yohe, Henry Jacoby, Richard Richels, and Benjamin Santer

January 5, 2021

Imagine a major climate change law passing the U.S. Congress unanimously? Don't bother. It turns out that you don't need to imagine it. Get this:

The Global Change Research Act of 1990 was passed unanimously[1] (100–0) in the United States Senate and by voice vote in the House of Representatives. Wow.

The law instructed all relevant federal agencies to intensify their separate research activities into climate change trends, impacts, and uncertainties and to coordinate their efforts under a newly created United States Global Change Research Program (USGCRP). Congress also recognized the need to communicate to the general public the societal and natural vulnerabilities derived directly or indirectly from current or projected climate change. To do that, the law mandated that the federal research community prepare regular national climate assessments (NCAs) to be distributed to the American people every four years by the sitting President.

The first assessment (NCA1) approved and released[2] in November of 2000, effectively began the communication process. It alerted Americans of growing threats posed by human-induced changes in local and regional climates.

Beginning around 2007, risk assessment became the accepted approach to understanding and communicating climate change impacts around the world. NCA3 in 2014 and NCA4 in 2018 therefore instructed writing teams[3] to characterize important climate change effects in terms of the two key principles of risk: the likelihoods of climate change impacts, and their consequences as measured by dollars, lives, other public health metrics, etc.

This risk-based framing meant that national assessments should report *high-risk possibilities* of all sorts: high-risk circumstances that could, for example, be (1) *highly likely to occur* with modest to moderate consequences; or (2) *more-likely-than-not to occur* with more serious consequences; and/or (3) *unlikely to occur* but with enormous and sometimes calamitous consequences.

By law, the incoming Biden Administration will be responsible for preparing the NCA5 for release in 2023, and the time is right for that assessment to advance to the next level by adding risk management to its organizational structure. The new USGCRP assessment team will clearly have to work within

---

[1] https://www.congress.gov/bill/101st-congress/senate-bill/169.

[2] https://www.globalchange.gov/browse/reports/national-assessment-potential-consequences-climate-variability-and-change-dvd.

[3] https://globalchange.gov/.

the already approved prospectus for NCA5.[4] But unless it pushes itself beyond the boundaries of past assessments, it will not focus stronger attention on how climate risks are being managed now and how they might be managed better in the future.

Moving beyond risk assessment to still more difficult questions .

Would that be a step forward? Clearly. Consider, for instance, what it could have added to the key findings[5] about increased vulnerability to coastal flooding reported in NCA4. They included several cautions:

- Vulnerability had been driven by human-induced sea level rise for decades, but it had not been evenly distributed along the nation's coastlines;
- The frequency of high-tide flooding had increased 5- to tenfold in some communities since 1965, but not in others;
- Flooding from extreme coastal storms like Nor'easters and hurricanes had generally made landfall with exaggerated storm surges and ponderous rainfall totals over short periods of time; and
- Adverse impacts from these types of storms are expected to increase as the planet warms over the next century.

NCA4 also estimated that highly cost-effective adaptation programs could reduce cumulative discounted future damages to coastal properties across the lower 48 states through 2100 by many billions of dollars, at the very least, and perhaps up to a few trillions of dollars along higher emissions futures.

A focus on risk management in upcoming NCA5 chapters could lead the authors to move beyond these assessments of risks and confront more difficult questions like: "What level of preparation at local, city, state, and regional levels would be required for investment in adaptation to achieve the damage avoidance earlier author teams had suggested would be feasible given changed conditions from those prevalent in 2018? And how might the federal government help (or hinder) in that regard?

NCA4 did present some examples of ongoing efforts to adapt to, mitigate, and provide relief from climate damages, but assessing capacities to *manage* risk could have brought more critical questions to the fore. For example:

- Do decision-makers across our federalist system work well together?
- Do they have and share the necessary information?
- Are their financial and human resources sufficient?
- Do bottlenecks or competing agendas impede efforts to reduce net damages?
- How can public understanding of and trust in climate action be improved?

Lessons learned from pandemic ... and meeting growing needs of courts .

---

[4] https://www.federalregister.gov/documents/2020/07/10/2020-14904/request-for-comment-on-the-draft-prospectus-of-the-fifth-national-climate-assessment.

[5] https://www.nationalacademies.org/our-work/americas-climate-choices.

Finally, consider what management lessons can be drawn from our challenging experiences in trying to manage the COVID-19 pandemic?

A substantial and growing number of insightful documents have been published, notably including the five volumes of America's Climate Choices.[6] Authors have described and dissected examples of success and of frustration in dealing with all sorts of external threats to human welfare: There is plenty of material to assess, integrate, and synthesize for the first time to help us fine-tune our capacities to attack climate change over a large and diverse country. Such an approach could lead to including risk management sections in sectoral and regional chapters of NCA5 with analyses that will inform and enhance an expanded adaptation chapter.

Increasing NCA5 attention to risk management is critical also for the judicial system. As climate effects multiply and management responsibilities grow, more cases involving the management of climate risks likely will arise in court dockets across the U.S. The Supreme Court has already charged judges[7] at every level "to determine whether proffered scientific testimony or evidence satisfies the standard of evidentiary reliability" because "a judge must ascertain whether it is ground[ed] in the methods and procedures of science."

Jurists frequently look to federally prepared scientific reports for guidance in this regard. They will certainly look more frequently to the NCA5 if its coverage of risk management practices provides insights into what might reasonably be expected of plaintiffs or defendants in cases involving climate risks.

Bringing risk management into the NCA is the next important step in its evolution in communicating climate vulnerabilities to the public. Doing so will illuminate what we know about incorporating the exploding knowledge of intensifying climate risk into public and private decision-making processes.

Afterword

The draft of NCA5 went out for review in summer 2022. The final report is unlikely to be made public until late 2023. Whether or not its assessment extends to specific federal measures to deal with climate risk, the report will be a useful input to risk management activities by state and city governments, and perhaps by firms in vulnerable industrial sectors. Many of these stakeholders face near-term risks of climate change, or even current risks. Cities, states, and affected industries are already dealing with climate risks by formulating policies and plans and reorganizing public agencies. This essay suggests that these risks need to be given greater consideration at the federal level.

---

[6] https://www.nationalacademies.org/our-work/americas-climate-choices.

[7] https://www.fjc.gov/sites/default/files/2015/SciMan3D01.pdf.

Essay 21. Deadlines Loom for Capitol Hill Action on Trump-Era Climate Issues

Prologue

Early in its term, the new Biden-Harris Administration and a new Congress faced a heavy schedule of policy development. Importantly, one of the opportunities to impose policy change had a short deadline and so could be easily overlooked in the early rush of the transition. Under U.S. law, Congress (with the signature of the President) can cancel a recently established federal regulation by a simple majority vote of both houses. However, this action must be taken within 60 legislative days (days the Congress is in session) after the regulation goes into effect. This provision offered the new Congress and Administration the opportunity to undo some of the more egregious midnight rule changes by the Trump Administration. We wanted to call attention to the fact that the clock was ticking.

This article first appeared as https://yaleclimateconnections.org/2021/01/commentary-deadlines-loom-for-capitol-hill-action-on-trump-era-climate-issues/.

Deadlines loom for Capitol Hill action on Trump-era climate issues

Gary Yohe, Henry Jacoby, Richard Richels, and Benjamin Santer

January 26, 2021

Much ink has been spilled in recent weeks, figuratively speaking, on what the Biden/Harris Administration's first 100 days in office reveal about its making climate change a top priority. Those words have flowed both at this[1] site[2] and many other venues.

The Washington Post's January 22 posting of "Tracking Biden's environmental actions"[3] is notable. Written by Post Pulitzer Prize winners Juliet Eilperin and Brady Dennis, with graphics editing by John Muyskens, the piece compiles Trump Administration environmental, conservation, and energy regulations and policies that the Biden team hopes to overturn or "unwind."

"Biden can overturn some of them with a stroke of a pen," they write. "Others will take years to undo, and some may never be reversed."

Listing 64 air quality and greenhouse gas initiatives, they count one (stepping back into the Paris Climate Agreement) as having been overturned and another 21 as being "easy" to reverse. They score another 27 Trump actions as "medium" – requiring rewriting a regulation or pursuing a successful court action; and 15 as "difficult" – requiring lengthy rule-making process, legislation, or involved court action.

Clock is running for action on eleventh-hour rules.

Along with the quickly-ticking clock in the White House on the administration's first 100 days, let's not lose sight of another clock that is also running on Capitol Hill. It is not necessarily identical, second-by-second, but just as relentless.

As is the case with so many others who have spent years working to bring climate change to the fore as a critical national and international issue, we are pleased with the day-one Biden Executive Order to re-enter the Paris Climate Agreement. Fortunately, given the din of opposition from the usual voices on Capitol Hill, that action does not require congressional approval. The U.S. will be officially back in on February 19th – and the global community can continue to welcome our return.

---

[1] https://yaleclimateconnections.org/2021/01/commentary-what-biden-and-democatic-senate-can-do-on-climate-in-their-first-100-days/.

[2] https://yaleclimateconnections.org/2021/01/ins-and-outs-of-congressional-review-act-and-climate-change-rules/.

[3] https://www.washingtonpost.com/graphics/2021/climate-environment/biden-climate-environment-actions/.

It's no surprise that the fawning Senate majority of the 116th Congress did not exercise its right under the Congressional Review Act (CRA) to un-do damaging Trump rules. But even with its razor-thin 50/50 split in the new Senate, the new 117th Congress is positioned to do so, without fear of having to overcome the 60-vote filibuster threshold. There is now, at least and at last, an opportunity during its first 60 legislative workdays for the Congress to conduct serious oversight of some of the especially offensive and glaring rules finalized in the waning days of the Trump Administration and, if it chooses, send nullifying legislation to the new President for signing.

A few especially egregious examples were published in the Federal Register literally at the 11th hour:

- A January 6, 2021, rule that EPA "give greater consideration to studies where the underlying response data" – frequently involving confidential human health issues protected by HIPAA law being made publicly available. That would knee-cap important health-based rulemaking on many toxic air pollutants and other contaminants as well as the mental health risks from climate change.
- A January 7, 2021, EPA rule on the threshold for "significant pollution contributions" that would end application of the Clean Air Act to many non-electric power plants that are sources of important greenhouse gases.
- A January 13, 2021, EPA rule that piggy-backed onto the January 7 rule to bar future greenhouse gas regulations from applying to oil refineries, manufacturing, plants, and other facilities.

Appropriate Senate and House committees of jurisdiction now owe it to their air-breathing constituents to seek expert analyses as they review these and other Trump-era health and safety rules before the clock runs out on them for such oversight; and, where appropriate, to take action.

In some cases, nullification under the Congressional Review Act may be most appropriate and most expeditious. In some other cases, Biden Administration executive action may be more appropriate. In still others, judicial action may be best.

The Trump Administration itself, like the 115th Congress taking office in January 2017, proved to be prodigious and, it must be acknowledged, an effective practitioner of the using the Congressional Review Act. Its actions nullified Obama/Biden Administration policies on climate change and other environmental and public health issues. That law, still untested in the courts, has been used only 17 times since it was enacted as part of the 1996 Newt Gingrich-inspired "Contract with America"; 16 of those came soon after the Trump Administration took office.

There's a saying familiar to us all: "What's good for the goose is good for the gander." It is especially apropos when the resulting actions enjoy bipartisan support from the public at large. This is a time when legislators should listen, and closely, to what their constituents are saying and expecting them to do.

Afterword

Ultimately, the new Administration and Congress applied the Congressional Review Act (CRA) to three last-minute rules made by the Trump Administration. Only one rule targeted energy and climate. In late 2020, the Trump Administration had gutted an Environmental Protection Agency rule regulating methane emissions from oil and gas facilities. By action under the CRA, the Trump regulation was voided. The rule was returned to its terms as established in the Obama Administration.

Essay 22. Biden's Executive Orders Have Broad Public Support

Prologue

Executive actions were a hallmark of the early days of the new Biden-Harris Administration. The new President wanted to hit the ground running, and it was expected that working through the Congress would be slow and challenging. Even in the days of a Trump Administration that was largely dismissive of the problem, there was a commitment to climate action by many cities, states, and industry groups around the nation. At the federal level, therefore, the new Administration was going with, not against, the tide of public opinion. We felt it was useful to highlight this fact, adding what weight we could in support of the executive initiatives of President Biden.

This article first appeared as https://yaleclimateconnections.org/2021/03/bidens-executive-orders-on-climate-have-broad-public-support/.

Biden's executive orders have broad public support

Gary Yohe, Henry Jacoby, Benjamin Santer, and Richard Richels

March 8, 2021

Governing from the White House by executive actions – whether by executive orders or variations thereon – has its pluses and minuses.

Executive orders, for instance, can help get past rigid partisan opposition and around the steep Senate filibuster requirements of at least 60 votes for passage. Some require action, but others are intended to signal the incumbent's perspectives and preferences.

Whatever form they take, however, they lack the gravitas and the staying power of actual legislation passed by both the House of Representatives and the Senate and signed by the president. So what's the point, given that they often are overturned by the next administration from the opposing party? It's a fair question.

Use of presidential powers independent of congressional approval continues under both parties, so there must be a good reason. In part, it's that those actions need not be – and they should not be – perceived simply as "one-off" pronouncements born of a transient political agenda. Instead, they can communicate support for policy actions that reflect societal, economic and/or cultural trends having significant and stable popular support.

Under President Trump, a number of executive actions[1] were intended to dismiss the scientific evidence about the causes and effects of human-caused global warming. Many were, in fact, undertaken with the express goal of reversing actions by the previous Obama-Biden Administration.

In stark contrast, the early Executive Office actions undertaken by the Biden/Harris Administration often point to undoing those Trump actions – a practice Washington Post White House and environmental reporter Juliet Eilperin has characterized as "the unraveling of the unraveling."

Rowing downstream surely beats rowing upstream

Unlike the situation during the Trump Administration, public survey data[2] continues to suggest broad public support for the Biden climate change actions, such as re-entering the Paris climate agreement and reviving more restric-tive regulations on greenhouse gas emissions from fossil fuel power plants. Metaphorically speaking, the Trump Administration was rowing upstream in many on its climate change stances against a clear and ever-growing tide of

---

[1] https://www.nationalgeographic.com/environment/2019/02/15-ways-trump-administration-imp acted-environment/.

[2] https://climatecommunication.yale.edu/visualizations-data/americans-climate-views/.

public support; the Biden Administration, in contrast, is enjoying the advantage of paddling downstream with that support strongly at its back.

In addition, one can reasonably argue that the currents of public opinion increasingly are in sync with private-market strategies that support positive climate action. Here are a few examples during Trump's term in office:

- The Trump Administration took office touting its plan to put coal miners back to work by reopening coal mines. Even early in his only term, though, energy companies were already closing[3] their existing coal-fired plants, and so miners' hopes for more and more jobs went unfulfilled.
- Buoyed by one of the provisions in the 2017 tax bill, the Trump Administration three years later took executive action to open the Alaska National Wildlife Refuge, ANWR, for drilling. Disappointingly for supporters of that action, the December 2020 auction[4] netted barely $144 million, not even 20% of what was expected and with most of it coming from the State of Alaska. Three major oil companies working in the state wouldn't bite on spending billions to find new oil they concluded has little market appeal, especially in the long run.
- The Trump Administration's EPA efforts to relax mileage and emission standards for vehicles and performance standards for oil companies were both greeted by resistance from many large automakers and oil companies who had already spent vast sums to comply with the more rigid Obama era rules. They had thereby created a competitive advantage[5] in their markets that they were not at all keen to see evaporate.

Now, contrast those examples with more recent experiences showing that the downstream current is growing stronger and more persistent. Here are a few more indicators:

- Legacy automakers like Ford and General Motors have joined[6] many others in investing heavily to move their fleets more quickly to electric power.
- A growing number of countries, states, and communities are committing themselves[7] to achieving net zero carbon emissions over the next few decades;
- Having recognized the potential of climate change risks and opportunities to shape significant reallocations of capital and thereby reconfigure the world's financial system, the world's largest assets manager, BlackRock Inc. has announced[8] that it will make accounting for climate change a central factor in its investment strategies.

---

[3] https://www.msn.com/en-us/money/companies/alliant-says-it-will-close-its-last-wisconsin-coal-fired-power-plant-by-2024-columbia-energy-center-has-operated-for-nearly-50-years/ar-BB1djREx.

[4] https://www.adn.com/business-economy/energy/2021/01/06/anwr-lease-sale-brings-in-144-million-in-bids-mostly-from-alaska-state-owned-corporation/.

[5] https://eelp.law.harvard.edu/2020/04/when-industry-support-for-stricter-regulation-is-good-business-considering-the-car-rules-and-methane-standards/.

[6] https://www.reuters.com/article/us-autoshow-detroit-ford-motor-idUSKBN1F30YZ.

[7] https://unfccc.int/news/commitments-to-net-zero-double-in-less-than-a-year.

[8] https://www.npr.org/2020/01/14/796252481/worlds-largest-asset-manager-puts-climate-at-the-center-of-its-investment-strate.

- The Federal Reserve Board in its November Financial Stability Report[9] introduced an entry on reporting climate risks as part of its financial stress testing procedures.
- Facebook has decided to add labels and links[10] to climate change postings with the goal of winnowing-out climate disinformation and elevating accurate climate coverage originating from widely trusted government and academic interests.

For a conventional start-of-year "What's In, and What's Out" column, it is now hard to find among 2021's "Ins" those rowing upstream against the current support for profits and jobs associated with clean energy. The heydays of buggy whips, iPods, single-edge shaving razors, and more are far behind us.

Instead, visions of higher economic values from employment-friendly CO_2 reduction efforts, cleaner and more secure renewable energy options, and a healthier environment are coming into clearer and clearer focus every week.

Put simply, while there for sure is more paddling ahead in the fight to manage climate change, we're at least now headed downstream – with the current and with the winds at our back.

Afterword

In its first few months, the Biden-Harris Administration released a flood of executive orders covering many different areas of federal policy. The early executive orders on climate change refocused federal agencies on an "all government" attack on the problem, putting climate at the center of U.S. foreign and domestic policy. Later executive orders ranged from an instruction to begin including climate risk in financial regulation to an order pausing new oil and gas leases on federal property.

Though they can be useful, executive orders are a weak instrument considering the scale of the transition needed to meet the climate threat. Presidential powers are limited, and policies established by executive order are not "durable" (as we argue in Essay 33 in Part VI) because they can be reversed by a subsequent President. In its first year, the Biden-Harris Administration was unable to gain congressional approval of climate actions. That situation has now changed with the passing in August 2022 of the Inflation Reduction Act (IRA). Among the many provisions of this Act are investments in building efficiency, public transit, strengthening of the electric grid

---

9 https://www.federalreserve.gov/publications/files/financial-stability-report-20201109.pdf.

10 https://www.upi.com/Top_News/US/2021/02/18/Facebook-will-soon-add-labels-links-to-posts-about-climate-change/6551613664198/.

and charging stations for electric cars, and greatly expanded federal financial support for carbon-free energy supplies and electric vehicles.

Perhaps more importantly, the IRA represents an acceptance at the federal level that climate action is accessible—and so sends a signal that taking account of climate risks and climate actions is in the best interests of American citizens. This signal, however, may still not be enough for widespread acceptance of the topic of our next essay—the direct pricing of carbon emissions.

Essay 23. There's a Simple Way to Green the Economy—And It Involves Cash Prizes for All

Prologue

There is a vast literature explaining the advantages of an emissions price in environmental policy, and dozens of countries have some form of penalty on CO_2 emissions—via either a cap-and-trade system or a tax. Three cap-and-trade systems are being employed in the U.S.—one is among eleven New England and Atlantic states, and separate systems are in place in California and Washington state. The Obama Administration proposed a national system which passed the House but failed in the Senate. Of late, cap-and-trade seems to have fallen out of favor among proponents of pricing, at least at the national level in the U.S., and current legislative proposals are for a national tax. As the new Biden-Harris Administration was forming its climate initiatives, it was an opportunity to restate the advantages of a tax on CO_2 emissions—and in particular, the potential uses of the revenue (the so-called carbon dividend). We hoped that might lead to serious consideration of an emissions tax by the Administration.

This article first appeared as https://www.theguardian.com/commentisfree/2021/jan/05/simple-way-green-economy-cash-prizes-carbon-dividend.

G. Yohe et al., *Responding to the Climate Threat*,
https://doi.org/10.1007/978-3-030-96372-9_23

There's a simple way to green the economy—and it involves cash prizes for all

Henry D. Jacoby

January 5, 2021

Over the past year – when societies around the world have had to grapple with their greatest challenge in decades – climate change hasn't been at the top of the agenda. But that doesn't mean it's gone away. Far from it – in fact, we just experienced the hottest September in 141 years,[1] and extreme warmth[2] recorded in the Arctic continues a disturbing trend. When the focus turns back to this ongoing existential threat, hopefully we'll have learned some lessons from the pandemic about what can be achieved when imaginative thinking is brought to bear.

Our approach towards tackling the climate crisis is necessarily going to be multipronged. But one powerful tool is that of a carbon tax. So far, however, only a few nations[3] have taken this route. Why?

First of all, how do taxes on carbon work? Basically, they penalize fossil fuels for the CO_2 emitted when they're burned, and in doing so offer a two-part

---

[1] https://www.climate.gov/news-features/features/september-2020-another-record-setting-month-global-heat.

[2] https://www.climate.gov/news-features/featured-images/2020-adds-another-year-extreme-warmth-warming-trend-arctic-ocean.

[3] https://carbonpricingdashboard.worldbank.org/.

advantage compared with other measures. They make non-polluting industries and products more competitive, and yield a flow of revenue that can be used to calm opposition to emissions reduction.

Weaning our economies off fossil energy involves making it less financially attractive. In market economies, most personal and business decisions are driven by prices, and wherever a fossil fuel is the cheapest source, and not forbidden, it will continue to dominate. Not only that, but fossil energy is a determined adversary, ploughing money back into research and development designed to push costs down so it can remain competitive, even as renewables become cheaper. A price penalty on fossil emissions counteracts this.

There are several ways to raise the prices of coal, oil and natural gas. For example, you can build a tax-and-trade system, which limits total emissions but encourages emitters to trade their carbon allowances. It's simpler, however, to just tax fossil fuels when they're burned, as it sends a clear price signal to the market, which a variable trading price doesn't. At the moment, taxes on fossil energy are collected across the supply chain, from the point of production, as with US state severance taxes, to final sale, as with gasoline taxes in many countries. It's messy.

For environmental effectiveness, and ease of collection, carbon taxes are best imposed at the earliest point you can: the wellhead or the mine mouth, the refinery output gate, or the port of entry for imports. That way, the incentive to reduce emissions spreads down through the economy. For example, a US tax of $50 per metric tonne of CO_2 would raise the price of oil leaving the Texas oil patch by about $21 a barrel, and increase prices throughout the country for motor fuel and products made using oil-based energy. This would percolate down to your local store: environmentally friendly goods would become relatively less expensive, and carbon-intensive ones would be pricier.

So, if taxes on carbon are so effective, why aren't they more widely used? Well, perhaps it's because of the associations we all have with the "T" word. Tax is when you take money away – from businesses, and once that feeds through into prices, from individuals. No one likes the idea of having less money. Then there are those who argue that adding taxes hurts the economy as a whole. Yes, this ignores the fact that any tax would be less damaging to GDP than the effects of climate change, which is having devastating impacts. But the short-termism built into the economic status quo makes that hard to appreciate.

Yes – no one really likes taxes. They're unpleasant to contemplate, and a hard sell, politically. But what if there was a way all of that could be neutralised? A small but imaginative policy tweak that rendered raising the price of CO_2 pollution not a tax, but a gift?

There are many ways to manage the proceeds from a carbon tax. It doesn't have to simply disappear into government coffers. And that's the secret: it's possible to design systems that achieve what is called revenue neutrality – where every dollar taken in tax is returned to people's pockets. One version of this idea would send the revenue to the public as a per-capita carbon dividend, in an annual check.

For example, in 2020 a $50 per metric tonne CO_2 tax would return each US household an annual dividend somewhere between $1,500 and $2,000. That's more than the pandemic stimulus checks distributed to most US taxpayers in light of the extreme economic situation. And yet it would come every year.

But what's the point of collecting a tax if you just give the proceeds back? It all comes down to incentives. The tax part of this arrangement would make carbon-intensive goods less attractive, and green ones more so. Environmentally friendly businesses would thrive. Polluting businesses would be incentivised to make their operations less damaging, driving green innovation in the process. Gradually, via the millions of consumer choices made every day, the economy would shift on to a more sustainable footing.

The dividend part would not only make millions of people happy – who doesn't like receiving a check in the mail? – it would have a social impact. Even when you factor in the increased cost of energy and other goods, all but the highest income groups – those who consume the most carbon-intense goods and services – would come out ahead, with the lowest income group benefiting most of all. This result should be especially welcome in the wake of the Covid-19 pandemic, which has imposed the harshest penalties on the least advantaged communities and cast a harsh light on underlying disparities in income and wealth.

There are other revenue-neutral designs, but they're not as good – one proposal, for example, involves a tax swap. Carbon tax[4] revenue could be used to lower a tax on labor, like the payroll tax. But this would be less favorable than a direct rebate to lower-income households. Cutting a corporate tax as part of a tax swap on the other hand, would favor wealthier income groups.

A carbon dividend feels like a novel, unusual idea. There certainly aren't many directly comparable fiscal mechanisms in place. But if now isn't the time to try bold new solutions – when we've seen that governments can move mountains in the right circumstances – then when is? And though it looks radical, the dividend really is just a rather elegant solution to a major problem, which neatly circumvents many of the usual political objections to increased taxation. It might even be the first highly popular tax.

Moving market-oriented economies off fossil energy is going to be a long and difficult struggle. Funds will also have to be found to ease the burden of the energy transition in fossil-dependent parts of the economy, helping displaced workers and supporting the communities where they live. But marshaling the power of the price system to rebalance the whole economy away from carbon-intensive industries – while supporting those on lower incomes – seems like a wonderful place to start.

Gary Yohe and Richard Richels contributed to the preparation of this article.

---

[4] https://www.theguardian.com/environment/carbon-tax.

Afterword

The Biden-Harris Administration did not propose a carbon tax in its first two years, not least because it didn't have the votes for it in the Senate. There was a possibility of implementing emissions control through existing authority under the U.S. Clean Air Act, which could have led to a carbon price through trading among the states. Unfortunately, that route now seems closed by a 2022 decision by the Supreme Court limiting the Administration's ability to apply the Clean Air Act without specific congressional approval.

Support for a carbon tax continues, however. In the current Congress and the one preceding it, nine bills have been filed that would establish a carbon "fee." Several of these bills have bipartisan sponsors. There is also substantial industry advocacy for this measure. For example, the Carbon Leadership Council—whose members include firms in the oil and gas, electric power and automobile industries—argues that the solution to the climate challenge is a carbon fee with the revenue returned to the pockets of the American people.

Essay 24. Biden Channels FDR on STEM Policy

Prologue

Restoring respect for science was a crucial task for the incoming Biden-Harris Administration. One way to do that was to elevate the role and prominence of the Office of Science and Technology Policy (OSTP). As part of the Executive Office of the President, OSTP advises the federal government on science and technology matters in domestic and international affairs. Largely moribund during the Trump years, its role and staff were reestablished, its director was raised to cabinet rank, and a top scientist was appointed to the post.

President Biden's letter appointing his new OSTP director drew a parallel between the need to reinvigorate science today and the challenge facing the country at the end of World War II. It cited FDR's appointment letter to Dr. Vannevar Bush, who was tapped to lead the post-World War II challenge to apply science and technology in meeting society's needs. The close "then and now" comparison of scientific and technical challenges offered us the opportunity to highlight the differences in political atmosphere in 1945 and in 2021. We also emphasized the importance of the Biden-Harris Administration's efforts to counter a pervasive public distrust of science—something that was not a problem in FDR's day.

This article first appeared as https://www.scientificamerican.com/article/biden-channels-fdr-on-stem-policy/?previewid=5C22384B-0948-422D-B7F0AE60D40A5F64.

Biden channels FDR on STEM policy

Henry Jacoby, Gary Yohe, Benjamin Santer, and Richard Richels

February 18, 2021

In November 1944, President Roosevelt wrote a letter to Vannevar Bush,[1] who was then director of the Office of Scientific Research and Development (OSDR). From there, Bush oversaw many of the scientific advances that contributed to victory in World War II. The end of the war was yet nine months away, but FDR asked Bush to advise him on how, "in the days of peace ahead," the experience of the OSRD could be applied for the betterment of society.

Today, pushing to end a global pandemic but facing a growing climate change crisis, President Biden has sent a similar letter to Eric Lander,[2] the incoming science adviser to the president and director of the White House Office of Science and Technology Policy (OSTP). In addition to advising the President on matters of science and technology, OSTP leads the coordination and implementation of science policies across the federal government.

The parallels between the FDR and Biden letters are striking. Both emphasize the role of science in health. FDR wanted guidance on how to continue the war of "science against disease" that was pursued during WWII. President Biden seeks information on the key "lessons learned" from COVID-19. How can the federal government be better prepared to protect public health during the next pandemic? The Biden letter also includes a complementary focus on the health of the planet: it asks breakthroughs in science and technology that can help address climate change.

The two letters also share a concern for the application of scientific research to the wider needs of society. FDR sought recommendations on how the government could promote the research of private and public organizations; 75 years later, President Biden asks for new ways to ensure that the U.S. is the world leader in the technologies and industries of the future.

Other commonalities in the letters relate to enhancing public understanding of science and to the challenge of nurturing and retaining top scientific talent. FDR wanted to "make known to the world" the contributions science had made to the United States—and could make in the future. Specifically, he wanted a program to discover and develop scientific talent. Today, the Biden letter seeks ways to preserve and strengthen the long-term health of U.S. science: by protecting scientific integrity, improving models of research support, bringing the brightest minds to bear on the scientific issues of our time, and encouraging scientists to serve in key roles in government.

---

[1] http://scarc.library.oregonstate.edu/coll/pauling/war/corr/sci13.006.4-roosevelt-bush-19441117-01.html.

[2] https://science.gmu.edu/news/letter-geneticist-eric-lander-president-elect-biden.

The Biden letter does contain one objective that is missing from FDR's list: ensuring that the fruits of science and technology are fully shared, not just among regions (an earlier concern), but across racial, gender and economic lines. This was a challenge coming out of the war, of course, but only in the intervening decades has it risen high on the national agenda.

Bush's report in response to FDR's request was entitled *Science: The Endless Frontier* . This influential and visionary document was submitted to President Truman in July 1945.[3] It called for a federal role in science—a role that continues today. Although President Lincoln had already created the U.S. National Academy of Science in 1863, and had fostered agricultural research through establishment of the U.S. Dept. of Agriculture in 1862, deeper federal involvement in directing and funding science was not common before WWII.

The bush report clearly articulated the need for broader federal involvement in science. One section of the report[4] was explicitly headed "Science is a Proper Concern of Government." In the wake of the report's release, federal support for science expanded far beyond agricultural research, importantly including the creation of the National Science Foundation.[5] Ultimately, *Science: The Endless Frontier* led to the embedding of science and technology research in many federal agencies.

The Biden letter asks for something very similar—calling for "general strategies, specific actions, and new structures" to meet the twin challenges of global pandemics and global climate change. In responding to this request, Lander may wish longingly for the political and cultural environment in Vannevar Bush's day:

- At the end of WWII, the U.S. was not just the global scientific leader (in part because of the talent that escaped from war-torn Europe). It was also the dominant economic power; much of Europe and Asia were in ruins.
- Working in the dominant scientific and economic country of the post-War period, U.S. scientists did not have to deal with a fear, all too common today, of being attacked for cooperative scientific work with colleagues in nations viewed as economic competitors.
- The Bush proposals were presented to a public that had greater trust in public institutions. There was no need to "protect scientific integrity within government."
- *Science: The Infinite Frontier* was received in Washington, D.C, at a time of relative political stability. For all but two of the next eight years (when the Senate switched), the same party held the White House and enjoyed substantial majorities in both houses of Congress.

Now consider the stark contrast facing Lander:

---

[3] https://www.nsf.gov/about/history/nsf50/vbush1945_summary.jsp.

[4] https://www.nsf.gov/about/history/nsf50/vbush1945.jsp.

[5] https://www.nsf.gov/about/history/nsf50/nsf8816.jsp.

- The U.S. has experienced a steep decline in public trust in government, a process that has been ongoing for at least the past 60 years.[6] More recently, similar drops in confidence have been found in other public institutions,[7] such as religious organizations, the medical establishment and news media.
- We now live in a time of "alternative facts," the wielding of willful ignorance as a political tool, and widespread belief in Qanon and other conspiracies. Moreover, misinformation and disinformation can now be rapidly disseminated to millions of citizens via the internet.
- Making matters still worse, a huge gap exists between the two major U.S. political parties in their trust in science and scientists. Crucial scientific issues—for example, the efficacy of different COVID-19 intervention strategies and the reality and seriousness of climate change—are viewed through powerful political lenses.[8]

The incoming Administration understands clearly the challenge of sustaining American science and directing its power to the frontiers of most urgent need. The president's intense focus on this task is evident from his public statements and from his selection of Lander—a distinguished life scientist, research leader and teacher—as the director of OSTP. Biden's unprecedented decision to elevate that office to Cabinet-level status also sends a clear signal: after four years in the wilderness, science matters again.

This is good news, but executive actions by President Biden will not be enough. Restoring trust in science, and using science to tackle the most pressing problems facing humanity, is a job for all of us, not for politicians or scientists alone.

FDR called for policy informed by data and scientific analysis, U.S. global leadership in science, and the use of science to improve public health. We echo those calls. We add to them the call to use science to shape a sustainable future for our planet. As we have learned during the COVID-19 pandemic, individual health depends on the health of our communities, our society, our scientific institutions and our government. Likewise, the health of our country is inextricably linked to the health of our planet's climate system. We will not be well if our planet is on life support.

Afterword

Over the decades following World War II, strong science capability was built up in many federal agencies. Unfortunately, faced with budget cuts and general disdain for their work under the Trump Administration, many federal scientists and technology experts abandoned government service. In addition to active efforts to recruit top talent and correct some of the budget problems,

---

[6] https://www.pewresearch.org/politics/2019/04/11/public-trust-in-government-1958-2019/.

[7] https://journals.sagepub.com/doi/abs/10.1177/0956797614545133.

[8] https://skepticalinquirer.org/2021/01/in-science-we-trust-twenty-country-pew-survey-shows-trust-in-scientists-with-major-caveats/.

the incoming Biden-Harris Administration took measures to enhance public confidence in science. For example, the OSTP created an interagency task force to review federal science policies and ensure that they "prevent improper political interference" from affecting research or data, and to guard against the "suppression or distortion of scientific or technological findings." The role of scientific expertise was increased in agency committees and advisory boards. One widely publicized example is that many industry appointees on the EPA's Scientific Advisory Board were replaced with experts on the relevant science.

Essay 25. A New Book Manages to Get Climate Science Badly Wrong

Prologue

In May 2021, physicist and former Undersecretary of Energy Steven Koonin released a book entitled *Unsettled: What Climate Science Tells Us, What It Doesn't, and Why It Matters*. We found many of his statements to be distracting, deflecting, or irrelevant. Given our risk-management perspective on how to plan and implement responses to climate change impacts, we were particularly concerned about the absence of well-defined uncertainty language such as that employed by the IPCC for its scientific assessments. As a result, it was difficult for a reader of "*Unsettled*" to tell which aspects of climate science are known with confidence and which are not.

The book attracted considerable attention in media outlets denying the reality and seriousness of climate change. Our essay was an effort to counter the book's failure to accurately reflect the current state of climate science and to highlight the value of a risk-focused approach to climate policy evaluation.

This essay first appeared as https://www.scientificamerican.com/article/a-new-book-manages-to-get-climate-science-badly-wrong/.

A new book manages to get climate science badly wrong

Gary Yohe

May 13, 2021

Steven Koonin, a former undersecretary for science of the Department of Energy in the Obama Administration, but more recently considered for an advisory post to Scott Pruitt when he was administrator of the Environmental Protection Agency, has published a new book. Released on May 4 and entitled *Unsettled: What Climate Science Tells Us, What It Doesn't, and Why It Matters*, its major theme is that the science about the Earth's climate is anything but settled. He argues that pundits and politicians and most of the population who feel otherwise are victims of what he has publicly called "consensus science."[1]

Koonin is wrong on both counts. The science is stronger than ever around findings that speak to the likelihood and consequences of climate impacts, and has been growing stronger for decades. In the early days of research, the uncertainty was wide; but with each subsequent step that uncertainty has narrowed or become better understood. This is how science works, and in the case of climate, the early indications detected and attributed in the 1980s and 1990s, have come true, over and over again and sooner than anticipated.

This is not to say that uncertainty is being eliminated, but decision makers have become more comfortable dealing with the inevitable residuals. They are using the best and most honest science to inform prospective investments in abatement (reducing greenhouse gas emissions to diminish the estimated likelihoods of dangerous climate change impacts) and adaptation (reducing vulnerabilities to diminish their current and projected consequences).

Koonin's intervention into the debate about what to do about climate risks seems to be designed to subvert this progress in all respects by making distracting, irrelevant, misguided, misleading and unqualified statements about supposed uncertainties that he thinks scientists have buried under the rug. Here, I consider a few early statements in his own words. They are taken verbatim from his introductory pages so he must want the reader to see them as relevant take-home findings from the entire book. They are evaluated briefly in their proper context, supported by findings documented in the latest report of the Intergovernmental Panel on Climate Change.[2] It is important to note that Koonin recognizes this source in his discussion of assessments, and even covers the foundations of the confidence and likelihood language embedded in its findings (specific references from the IPCC report are presented in brackets).

Two such statements by Koonin followed the simple preamble "For example, both the literature and government reports that summarize and assess the state of climate science say clearly that...":

---

[1] https://tambonthongchai.com/2021/03/22/steve-koonin-lecture/.
[2] https://www.ipcc.ch/site/assets/uploads/2018/02/WGIIAR5-Chap19_FINAL.pdf.

- "Heat waves in the US are now *no more common* than they were in 1900, and that the warmest temperatures in the US have not risen in the past fifty years." (Italics in the original.) This is a questionable statement depending on the definition of "heat wave," and so it is really uninformative. Heat waves are poor indicators of heat stress. Whether or not they are becoming more frequent, they have clearly become hotter and longer over the past few decades while populations have grown more vulnerable in large measure because they are, on average, older [Sect. 19 .6.2.1]. Moreover, during these longer extreme heat events, it is nighttime temperatures that are increasing most. As a result, people never get relief from insufferable heat and more of them are at risk of dying.[3]
- "The warmest temperatures in the US have not risen in the past fifty years." According to what measure? Highest annual global averages? Absolutely not. That the planet has warmed since the industrial revolution is unequivocal with more than 30 percent of that warming having occurred over the last 25 years, and the hottest annual temperatures in that history have followed suit [Section SPM.1].

Here are a few more statements from Koonin's first two pages under the introduction that "Here are three more that might surprise you, drawn from recently published research or the latest assessments of climate science published by the US government and the UN":

- "Greenland's ice sheet isn't shrinking any more rapidly today than it was eighty years ago." For a risk-based approach to climate discussions about what we "should do," this statement is irrelevant. It is the future that worries us. Observations from 11 satellite missions monitoring the Arctic and Antarctic show that ice sheets are losing mass six times faster[4] than they were in the 1990s. Is this the beginning of a new trend? Perhaps. The settled state of the science for those who have adopted a risk management approach is that this is a high-risk possibility (huge consequences) that should be taken seriously and examined more completely. This is even more important because, even without those contributions to the historical trend that is accelerating, rising sea levels will continue to exaggerate coastal exposure by dramatically shrinking the return times of all variety of storms [Sect. 19 .6.2.1]; that is, 1-in-100 year storms become 1-in-50 year events, and 1-in-50 year storms become 1-in-10 year events and eventually nearly annual facts of life.
- "The net economic impact of human-induced climate change will be minimal through at least the end of this century." It is unconscionable to make a statement like this, and not just because the adjective "minimal" is not at all informative. It is unsupportable without qualification because aggregate estimates are so woefully incomplete [Sect. 19 .6.3.5]. Nonetheless, Swiss Re recently released a big report on climate change[5] saying that insurance companies are underinsuring against rising climate risks that are rising now and projected to continue to do so over the near term. Despite the

---

[3] https://www.who.int/news-room/fact-sheets/detail/climate-change-heat-and-health.

[4] https://climate.nasa.gov/news/2958/greenland-antarctica-melting-six-times-faster-than-in-the-1990s/?ftag=YHF4eb9d17.

[5] https://corporatesolutions.swissre.com/dam/jcr:071c5173-84d0-43d2-8038-0e287e6f3306/climate-change-report.pdf.

uncertainty, they see an imminent source of risk, and are not waiting until projections of the end of the century clear up to respond.

The first of these misdirection statements about Greenland is even more troubling because the rise in global mean sea level has accelerated. This is widely known despite claims to the contrary in Chapter 8 which is described in the introduction as a "levelheaded look at sea levels, which have been rising over the past many millennia." Koonin continues: "We'll untangle what we really know about human influences on the current rate of rise (about one foot per century) and explain why it's very hard to believe that surging seas will drown the coasts any time soon."

The trouble is that while seas have risen eight to nine inches since 1880, more than 30 percent of that increase[6] has occurred during the last two decades: 30 percent of the historical record over the past 14 percent of the time series. This is why rising sea levels are expected with very high confidence to exaggerate coastal exposure and economic consequences [Section 19 .6.2.1].

His teaser for Chapter 7 is an equally troubling misdirection. He promises to highlight "some points likely to surprise anyone who follows the news—for instance, that the global area burned by fires each year has declined by 25 percent since observations began in 1998." Global statistics are meaningless in this context. Wildfires (if that is what he is talking about) are local events whose regional patterns of intensity and frequency fit well into risk-based calibrations because they are increasing in many locations. Take, for example, the 2020 experience. Record wildfires[7] were seen across the western United States, Siberia, Indonesia and Australia (extending from 2019) to name a few major locations.

Take a more specific example. From August through October of 2020, California suffered through what became the largest wildfire in California history. It was accompanied by the third, fourth, fifth and sixth largest conflagrations in the state's history[8]; and all five of them were still burning on October 3. Their incredible intensity and coincidence can only be explained by the confluence of four climate change consequences that have been attributed to climate changes so far: record numbers of nighttime dry lightning strikes during a long and record-setting drought, a record-setting heat wave extending from July through August, a decade of bark-beetle infestation that killed 85 percent of the trees across enormous tracks of forests, and long-term warming that has extended the fire season by 75 days.

So, what is the takeaway message? Regardless of what Koonin has written in his new book, the science is clear, and the consensus is incredibly wide. Scientists are generating and reporting data with more and more specificity about climate impacts and surrounding uncertainties all the time. This is particularly

---

[6] https://www.climate.gov/news-features/understanding-climate/climate-change-global-sea-level.

[7] https://www.resilience.org/stories/2020-10-27/world-on-fire-2020-experts-explain-the-global-wildfire-crisis/.

[8] https://yaleclimateconnections.org/2020/10/multiple-extreme-climate-events-can-combine-to-produce-catastrophic-damages/.

true with regard to the exaggerated natural, social and economic risks associated with climate extremes—the low-probability, high-consequence events that are such a vital part of effective risk management. This is not an unsettled state of affairs. It is living inside a moving picture of what is happening portrayed with sharper clarity and more detail with every new peer-reviewed paper.

Afterword

Dr. Koonin continued to pitch his book well past the fall of 2021. In late December, "*Unsettled*" was featured in a *Forbes* book talk where Dr. Koonin expressed hope that the attention he had drawn to climate change (book sales had totaled 100,000) would "prompt greater scrutiny of the 'alleged climate crisis'." However, his repeated proposals for an adversarial red team/blue team debate focused on the credibility of climate change science have not (as yet) received any takers. The on-the-ground reality of climate change is simply difficult to reconcile with Koonin's thesis that climate change is not a big deal.

Essay 26. This Is How Extreme Weather Events and Climate Change Are Connected

Prologue

The with each of 2021's extreme and damaging weather events came a common question: "Any connection with climate change?" This provided an opportunity to explain what is known in climate jargon as the "event attribution" question: Can individual extreme events be tied directly to climate change? And how can one explain the connections between the two without subjecting the reader to perhaps unfamiliar terminology of probability and statistics?

One way to do this is by using a metaphor familiar to all to illustrate how climate change can increase the likelihood of extreme events. Throwing two dice in a game of Monopoly came to mind. Assume the dice throws yield numbers like 10, 11 and 12 far more often than would be expected by pure chance. It's possible that the dice are loaded, or it may just be chance. But if a 14 or 16 shows up, you know the outcomes from the dice have changed fundamentally. We're now playing a game much more serious than the monopoly we grew up with as kids. We feel like Mother Nature has been rolling some 15 s, and we wanted to explain how this is happening with help from human-caused warming.

This essay first appeared as https://thehill.com/opinion/energy-environment/564933-this-is-how-extreme-weather-events-and-climate-change-are.

This is how extreme weather events and climate change are connected

Gary Yohe

July 26, 2021

It is widely known that weather is not climate. It is also widely known that climate, and therefore climate change, can significantly affect the weather over time — not only with respect to changes in trends in important indicators like temperature and precipitation, but also with respect to the likelihoods and intensities of extreme events.[1]

The correlation between climate and impacts is, however, most often complicated by confounding factors that include all variety of human behaviors; and that is why something called "forensic attribution" is now a growth industry. Benjamin Santer crafted an up-to-date and extensive review[2] of the state of knowledge of forensic attribution even in its formative stages and summarized in a recent op-ed.[3]

In 2021, when an extreme event like the recent flash flooding in western Germany occurs, the question is "How much of the observed extreme intensity can be attributed to climate change?"

As is always the case in climate change, the answer to the question is always "It depends." But on what? Is there any way to sort out the facts about both change and the rate of change?

It turns out that a scientifically rigorous answer to this question is "Yes." Since responding to climate change is a risk management problem, the world has accepted that the key to organize our thoughts must focus on the two fundamental components of risk — likelihood and consequence or, in the vernacular, frequency and intensity.

Take, for example, the very recent severe flooding in Western Germany.[4] Three months of rain fell in three hours. Flooding caused enormous damage with nearly 200 deaths and more than 300 missing as of July 20. It was the worst flooding event in more than 50 years — but that means that there had been a similar flooding event in the lifetimes of some residents. It was very unusual to see it repeat in 2021, but it was not statistically difficult to explain.

This episode is like playing monopoly when somebody rolls 12 three times in a row — weird, but possible (0.0002148 percent probability). You might suspect that the dice had been loaded to produce high numbers, but the chance was

---

[1] https://www.nasa.gov/mission_pages/noaa-n/climate/climate_weather.html.

[2] https://www.nasa.gov/mission_pages/noaa-n/climate/climate_weather.html.

[3] https://thehill.com/opinion/energy-environment/564722-how-do-climate-scientists-study-the-causes-of-climate-change.

[4] https://www.scotsman.com/news/world/where-are-the-floods-in-germany-map-of-country-shows-worst-affected-areas-in-2021-flooding-disaster-3310455.

not zero that it was just a rare event. In other words, nobody could claim without further evidence that the severity of the flooding was caused by climate change. When all is said and done, the enormous damage and the loss of life will likely be attributed to human behaviors and development decisions made over the past half-century since the last flood of this magnitude occurred.

But now, take as another example, the enormous heatwaves in the western portion of the United States. Over recorded history, there had been only three other days with temperatures that reached maxima of 106[5] degrees Fahrenheit in Portland, Oregon (July 2, 1942; July 30, 1965; and Aug. 10, 1981).

Then, residents experienced three successive days above 106 degrees born during the end of June of a heat dome generated by a wildly variant jet stream trajectory.

As a third example, take Hurricane Harvey in 2017 dropping 42 inches of rain over Houston over the course of 36 hours. It created the third 500-year flood in four years. Also recall that Hurricane Dorian in 2019 stalled over Nassau in the Bahama for more than a day with similar effect. Climate change does not cause hurricanes; but Harvey's behavior and Dorian's behavior and Maria's devastating swath through Puerto Rico in 2017 were typical of recent hurricanes — stalling over land because their steering currents had been weakened and so they had no place to go. Why? Likely because of shifts in the jet stream[6] and other atmospheric currents had been caused by planetary warming that diminished the energy gradient between the poles and the equator.

To return to the monopoly metaphor: These events cannot be explained by rolling ordinary dice with unusual results. It is more like rolling 16 three or four times in a row, and therein lies the irrefutable evidence that the game has changed. In the old game, 16 is an impossible number to roll — and it's happening more than once is still ridiculous in the new one. In the climate context, we have nearly irrefutable evidence that Mother Nature has changed the dice because, in cases like these, there is simply no other explanation.

These have all been stories about likelihood. But what about intensity (consequence)? Consider one more example. Only three of the state's largest 20 fires (in terms of acres burned) had burned prior to 2000 — but nine of the biggest 10 had occurred since 2012.

Extremes were becoming more likely and growing larger. The year 2017 saw 9,270 fires burn a then record of 1.5 million acres. The Mendocino Complex fire the next year became the "largest wildfire in California history."

Things were not all that unusual until 2020 when the historical mold exploded. The Complex fire would become the new largest fire in California history. It

---

[5] https://www.pdxmonthly.com/news-and-city-life/2021/06/what-was-portland-up-to-the-last-times-it-got-this-hot.

[6] https://theglobepost.com/2019/10/22/lessons-hurricane-dorian/.

started in August. Soon thereafter, four more fires would become the third, fourth, fifth and sixth largest in history. By the beginning of October, they were all still burning with only minimal control having been established.

Wildfires have long been a natural part of the forest environment in California, of course. As time passed, damage to life and property increased markedly as more people moved into vulnerable forested areas. Changes in forest management contributed, as well. Fire suppression policies on federal land reduced the brush-clearing value of deliberately setting control blazes. But none of these or other non-climate causes of increased fire danger have increased sufficiently over the five years to account for the erupting devastation and risk. Something else is required to explain what has been happening, and that something is climate change effecting consequence if not likelihood

Many of the fires in 2020 were caused by dry lightning, but their points of ignition appeared in the midst of a record drought.[7] Historic heat had sustained massive infestations of bark beetles over many years resulting in large stands of dead trees covering more than 85 percent of nearly every forest in harm's way. And the fire season had grown by 75 days a year.

The keys to quick attribution for the informed and engaged citizen are two in number – something that changed the likelihood of an extreme event quickly and/or something that changed its consequences *even more* quickly.

Afterword

Each passing year sees more examples of extreme events that could have been used in making the points contained in this essay. Many of these events are causing serious economic and environmental consequences and loss of life. People experiencing these damages are considering how to seek redress from those who might be blamed for causing the problem. A number of U.S. cities and states are now seeking compensation from oil companies for damages related to climate change.

There are many issues involved in making a legally sound case in this type of litigation. One set of these issues concerns how to establish the relationship between changes in the climate and a particular local extreme event, and how to quantify the resulting damage. Making those connections is important for other reasons as well. For example, a causal link between a changing climate and damage-causing extreme events is a basis for seeking aid under the Loss and Damage article of the Paris Agreement.

---

[7] https://yaleclimateconnections.org/2020/10/multiple-extreme-climate-events-can-combine-to-pro duce-catastrophic-damages/.

Essay 27. "Never Before" (NB4) Extreme Weather Events…. and Near Misses

Prologue

The year 2021 produced its share of off-the-charts extreme events. Some extreme heat events showed up in surprising places—like Seattle and Portland—so they generated greater than usual press coverage. The public focus on these events provided us with an opportunity to tie them together and provide a larger picture of ongoing change and rising risk. In the resulting essay, we integrated points explored in earlier essays (particularly in Essay 26) and added an acronym (NB4) in an effort to draw attention to the extraordinary events we are witnessing. More complete representation of rising climate risk at the "tails" of distributions (like distributions of maximum daily temperatures) requires application of advanced statistical methods. But Mother Nature, aided and abetted by human-caused warming, has begun to help us out with the statistics. One could argue, with some justification, that it's getting easier to evaluate tail risks because each year is bringing more samples of off-the-charts extreme events.

This essay first appeared as https://yaleclimateconnections.org/2021/09/never-before-nb4-extreme-weather-events-and-near-misses/.

'Never before' (NB4) extreme weather events.... and near misses

Gary Yohe

September 9, 2021

A recurring and troubling pattern of first-time historic weather events provides firm support for citizen and leaders to acknowledge human causation and take needed mitigation and adaptation steps.

New Jersey Governor Phil Murphy visits storm-ravaged Mullica Hill on September 2, viewing damages caused by 'remnants' of Hurricane Ida. (Photo credit: Edwin J. Torres/NJ Governor's Office).

Attributing extreme events to climate change – including those highly reported though the media – is a difficult task frequently requiring lots of time to complete rigorously. The usual mantra is that climate change did not cause X, but climate change did contribute significantly to its intensity and/or its frequency. Which raises the question: "By how much?"

But experience on the ground sometimes makes that attribution to climate change a no brainer. How so? Because no other influence can explain many of the recent events because there is no precedent for their having ever been happened before. Call them "Never Before" in history events (NB4s).

The mundane "Who cares?" version of an NB4 event can be found in the time series of an index of annual mean surface temperature. The five-year trend comparison has been *de rigueur* for decades, but over just the past 20 years, the "This has been the hottest year ever" framing[1] has been assigned to five of those years.

---

[1] https://climate.nasa.gov/vital-signs/global-temperature/.

Another example of a time series worrisome to many experts involves Hurricane Harvey, in 2017. Harvey stalled over Houston for nearly two days. It dropped 42 inches of rain while it was just hanging around with nowhere to go. Stalling of hurricanes has been attributed[2] to a reduced temperature difference between the poles and the tropics. It is a signature of climate change that now includes Ida over Louisiana. In Houston, climate change caused the third "500-year flooding" event in four years – certainly a damaging NB4.

In the summer of 2020, leaking methane from the melting permafrost across tundra in Siberia released methane that spontaneously ignited when temperatures[3] well above the Arctic Circle exceeded 100°F. The high temperatures are a product of global warming, but the interaction with the tundra is a very troubling NB4.

Hurricane Ida was the second Category 4 (nearly a Cat 5) storm to make landfall in Louisiana in two years. Ida tied the record[4] for gaining intensity when approaching landfall. The cause of that rapid intensification? Temperature of the Gulf of Mexico waters provided fuel to buttress the intensity. Those water temperatures across the Gulf ranged between 88°F and 90°F *to* a depth of 150 feet – never before in recorded history.

Subsequently, how is it possible that more than 15 times as many people died from exposure to Ida in eight mid-Atlantic states than in Mississippi and Louisiana combined? Because the severity[5] was unexpected, and many people were unprepared.

In New York City, sustained rain for one hour exceeded three inches during Hurricane Henri in early August, an all-time record. Less than two weeks later, the remnants of Ida piled on[6] with a new all-time record of 3.15 inches for New York City and 3.24 inches for Newark, New Jersey. Surely another NB4, and especially for piling on. IDA was an NB4 event at least three times over.

Who should care? Surely insurance companies should ... and do. They diversify by geography against severe storm events. They increasingly face storm liabilities not only in the anticipated urban and rural and coastal areas along the Gulf of Mexico, but also, and increasingly, in the more densely populated broadly distributed areas of New York City, New Jersey, and even Philadelphia. The former they've anticipated. The latter, not so much.

---

[2] https://yaleclimateconnections.org/2020/10/multiple-extreme-climate-events-can-combine-to-pro duce-catastrophic-damages/.

[3] https://weather.com/news/climate/news/2020-06-21-siberia-russia-100-degrees-heat-record-arctic#:~: text=Siberia%2C%20one%20of%20the%20world%27s%20coldest%20places%20in,%28F%29%20t his%20year%20before%20Dallas%20or%20Houston%20did.

[4] https://search.yahoo.com/yhs/search?hspart=tro&hsimp=yhs-freshy&grd=1&action=nt&type= Y219_F163_204671_012221&p=ida+increase+in+intensity+before+landfall.

[5] https://edition.cnn.com/us/live-news/ida-aftermath-09-02-21/h_6553ab3339b8a7aacf8617b7d08b 2dac.

[6] https://edition.cnn.com/us/live-news/ida-aftermath-09-02-21/h_6553ab3339b8a7aacf8617b7d08b 2dac.

And then, not to be outdone or forgotten, there are the rampant wildfires[7] in California: 2018 brought the largest fire in Cal Fire's recorded history. The following year, 2019, was more modest in its aggression, but 2020 erupted with a new largest fire in history. The conflagration was also burning at the *very same* time as the 3rd, 4th, 5th and 6th largest fires in history. Why the intensity? Megadrought, pine bark beetles that had not suffered through their usual winter freeze for a decade, and extreme record heat combined with record dry lightning. 2020 was an NB4 year.

This calendar year, 2021, has shown no sign of backing down from the challenge to be the worst. It, too, boasts an NB4 claim not only from the same causes, but also for a different reason: No California fire in history had ever climbed the Sierra Nevada mountains and rolled down the eastern side toward Nevada. The Dixie fire accomplished that heretofore-unprecedented feat. But wait, as the cheap cable commercials say, there's more: A month or so later, the Caldor fire did the same thing, soon seriously threatening South Lake Tahoe for the first time in history. Consider it an NB4 two-fer.

With regard to heat waves,[8] look across the U.S. Pacific Northwest and western Canada. Seattle, for instance, experienced three successive days in the summer of 2021 with maximum temperatures of more than 100°F (June 26–28, 2021). In all of prior recorded history, Seattle had seen only three days above 100°F (July 16, 1941; July 20, 1994; and July 29, 2009). Portland, Oregon, and other areas – places where residential air conditioning are few and far between – fared no better and in some places worse.

And then there is rain in Greenland for the first time, the biggest tornado (spawned by Ida) in New Jersey history, seven inches of rain in Central Park tying the 1927 record, and so on …

It is time for the Congress and its citizen constituents, decision-makers of all sort, and opinion-makers of all political persuasions to acknowledge that human-driven climate change is undeniably causing catastrophic effects in ways never seen before. And those often-calamitous effects are not only in the usual suspect places and the results of predictable reasons.

They are occurring unpredictably and in surprising and unexpecting, and therefore often least prepared, places.

Afterword

A few months after publication of this essay, Hurricane Larry made landfall as the first Category 1 storm to strike Newfoundland since 1775. The drought in the western U.S. continued, showing unprecedented persistence for nearly two decades. The rivers around the north side of Yellowstone

---

[7] https://www.fire.ca.gov/media/4jandlhh/top20_acres.pdf.

[8] https://www.seattletimes.com/seattle-news/3-of-seattles-4-hottest-days-on-record-just-happened/.

National Park suffered historic flooding in June 2022. That June also saw the hottest temperature on record by multiple degrees in Texas; the heat emergency continued well into July. The U.K. broke temperature records by nearly 10% on a single day in July.

And so it goes. The growing accumulation of these and other NB4s cannot be denied. Are they the "new normal?" Absolutely not. They are alarming data points along the new trajectory that our planet's climate system is following— a trajectory that human actions have influenced. We are headed for worse unless we change our ways. Humans can, though, alter the trajectory and can slow or even stop the pace of change.

Part VI

The Ongoing Challenge

This final set of essays was written after the Biden-Harris Administration and the 117th Congress were well under way. The first (Essay 28) highlights a 2021 event that helped quash any doubts about the need for climate action—describing as it does the long process that led the IPCC to confidently attribute the ongoing global warming to human influence. The remaining essays in this Part focus on the ability of the U.S. and other nations to actually take the needed action to address climate change.

Three of these essays discuss some of the most serious external events and long-standing stresses that hinder progress. One essay (29) is concerned with the growing geopolitical tension and possible fragmentation of the post-World War II international order, threatening collaboration among nations on science and policy analysis. It also addresses the way the Ukraine war and scramble to replace Russian energy are disrupting plans for an orderly transition away from fossil fuels. A related essay (30) considers the possible salutary effect of the urgent effort to shift off fossil fuels on the struggle between democracy and autocracy. The last of this trio of essays (31) rehashes the long-standing disagreement between rich and poor countries over who should do what, an unresolved problem that drags on in international climate negotiations.

Two final essays concern domestic U.S. developments. In one (32) we express worries about possible misuse of some of the terms publicly used to categorize climate risk—behavior that could undermine public trust in science. And in the other (33) we call out the fact that durability is an essential feature of the policies required to deal with a long-term challenge like climate change.

Essay 28. How IPCC Went from "Not Proven" that We Cause Climate Change in 1990 to "We Are Guilty" in 2021

Prologue

In October 2021, Working Group I of the Intergovernmental Panel on Climate Change released its report to the IPCC's Sixth Assessment of climate change science. The report received considerable public attention, in part due to its strong statement on the human contribution to climate change. We produced a brief history of this more confident language because it is a nice lesson about the dynamic nature of science. Through a series of IPCC assessments, understanding of the human role in climate change advanced dramatically—progress that was enabled by better and longer climate data records, improved computer models, and more insightful analysis methods.

This essay first appeared as https://thehill.com/opinion/energy-environment/567678-how-ipcc-went-from-not-proven-that-we-cause-climate-change-in-1990.

G. Yohe et al., *Responding to the Climate Threat*, https://doi.org/10.1007/978-3-030-96372-9_28

How IPCC went from 'not proven' that we cause climate change in 1990 to 'we are guilty' in 2021

Benjamin Santer

August 12, 2021

The Intergovernmental Panel on Climate Change (IPCC)[1] advises the world's governments on climate change science, impacts and response strategies, and recently released their latest report on climate change science. It was the sixth in a series of scientific assessments that began in 1990.

Much has been written[2] about the Sixth Assessment Report (AR6). It attracted world-wide attention for its finding of "unequivocal"[3] human influence on the climate system. The scientific jury comprised hundreds of experts in climate science. They reached a clear verdict: Humanity bears primary responsibility for the global warming of roughly 1.1 degrees Celsius since 1850.[4]

Most commentators on AR6 view the report as a somber warning.[5] A warning that planetary scale warming is already leading to more intense heat waves, droughts and floods.[6] A warning that global warming is already driving sea-level rise, an existential threat[7] to small island nations and millions living in low-lying coastal areas.[8] A warning that all countries need to act swiftly and decisively[9] to reduce emissions of greenhouse gases from fossil fuel burning. A warning that if we lack the political will to act, the forecast is poor for the well-being[10] of present and future citizens of this planet.

Back in 1990, the first IPCC report[11] concluded that the scientific jury was still out on the identification of human-caused global warming. Their finding? "The unequivocal detection of the enhanced greenhouse effect from observations is not likely for a decade or more." The 1990 verdict of "not proven" has now been replaced by a very different verdict in 2021: unequivocal human influence on global climate.

---

[1] https://www.ipcc.ch/.

[2] https://www.washingtonpost.com/climate-environment/2021/08/09/ipcc-climate-report-global-warming-greenhouse-gas-effect/.

[3] https://www.ipcc.ch/report/ar6/wg1/downloads/report/IPCC_AR6_WGI_SPM.pdf.

[4] https://inews.co.uk/news/what-intergovernmental-panel-climate-change-ipcc-report-climate-change-global-warming-1141998.

[5] https://www.nytimes.com/2021/08/09/climate/climate-change-report-ipcc-un.html.

[6] https://www.theverge.com/2021/8/9/22613531/climate-change-united-nations-report-extreme-weather-ipcc.

[7] https://www.independent.co.uk/climate-change/news/sea-level-rise-ipcc-report-2021-b1899177.html.

[8] https://www.cnn.com/2021/08/01/africa/lagos-sinking-floods-climate-change-intl-cmd/index.html.

[9] https://news.un.org/en/story/2021/08/1097362.

[10] https://www.usnews.com/news/health-news/articles/2021-07-08/climate-change-already-causes-5-million-extra-deaths-per-year.

[11] https://www.ipcc.ch/report/ar1/wg1/.

Why did the verdict change?

I study the causes of climate change,[12] and participated in each of the IPCC's six climate science reports. This gives me an informed perspective on how "not proven" in 1990 became "humans are guilty" in 2021.

Many factors enabled this scientific progress. One factor was better, longer and more complete observations of climate change.[13] We now study Earth's climate from space, the atmosphere, the oceans and land surface. A wide variety of measurement platforms probe the climate system in amazing detail. Global changes in temperature, moisture, sea level and ice cover are routinely monitored. Less-familiar properties are also measured, like the liquid water content of clouds, the mass balance[14] of major ice sheets and variations in ocean color driven by phytoplankton blooms.[15]

Collectively, this rich set of observations provides crucial information for understanding causation. It's tough to understand why climate is changing if you don't know how it's changing.

Another key factor shaping the trajectory from IPCC's first 1990 report (AR1) to the sixth released this week was the development of better computer models of the climate system.

Einstein was fond of "Gedankenexperimente" — thought experiments[16] you could perform in your head to shed light on important physical principles. He would have loved climate models. They are the ultimate thought experiment. A modeler can change different influences on climate, either individually or in concert. This makes it easier to identify the unique fingerprints that are caused by different human and natural influences on climate.

At the time of IPCC's 1990 report, a half-dozen climate modeling groups existed. In 2021, the number of climate models assessed had swelled by an order of magnitude. The complexity of models also increased along the scientific road to the latest report. The ocean's role in the climate system was represented more realistically. Aspects of the climate system left out of the 1990 models were gradually added: the uptake and release of carbon from the oceans, soil and vegetation. Atmospheric chemistry, marine biogeochemistry and ice sheets.

Each addition of a new component of the climate system allowed scientists to study key "interaction terms." Many of these interactions are critically important. We need to know the impact of global warming on the ocean's ability to

---

[12] https://www.sigmaxi.org/Error?aspxerrorpath=/news/keyed-in/post/keyed-in/2019/11/11/how-do-we-know-that-human-activities-have-affected-global-climate.

[13] https://www.nature.com/articles/s41558-019-0424-x.

[14] https://www.nature.com/articles/s41586-019-1855-2.

[15] https://www.sciencedaily.com/releases/2020/03/200319125151.htm.

[16] https://en.wikipedia.org/wiki/Einstein%27s_thought_experiments.

absorb the CO_2[17] emitted from fossil fuel burning. We need to understand how ocean warming and sea-level rise affect the stability of massive ice shelves[18] off the coast of Antarctica. We need to know how the melting of Arctic sea ice and the Greenland ice sheet might influence ocean circulation[19] in the North Atlantic.

The availability of more and improved climate models had many added benefits. Each model has a different representation of the physics, chemistry and biology operating in the real-world climate system. By analyzing a large collection of models, scientists could quantify uncertainties in projections of 21st century climate change. They could assess whether detection of human fingerprints in observations was robust to model uncertainties.[20] In the robust results, they could try to understand the physical processes driving common changes in climate.

Better observations and models are only part of the story of scientific progress from AR1 to AR6. Progress has many parents. There are now more rigorous protocols for evaluating and testing climate models.[21] There is greater computational power — a necessary condition for running increasingly more complex models with finer spatial granularity. There are better platforms[22] for sharing and analyzing enormous volumes of observational data and climate model output. From ice cores, tree rings, corals, sediment cores and many other paleoclimate records, there is improved understanding of climate changes over deep time, showing that global warming since the Industrial Revolution is unusually large and rapid.[23]

And there is now a larger community of climate scientists — a community driven by a common desire to understand how and why our atmosphere and oceans are changing, and what those changes portend for the future.

Progress has also been made by responding to those who have kicked the tires of climate science. Let me give an example. Back in 1995, I was the convening lead author of chapter 8 of the IPCC AR2[24] report. Our chapter concluded that "the balance of evidence suggests a discernible human influence on global climate." Many folks kicked that tire. They claimed satellite data showed "no warming whatsoever"[25] of the lowermost layer of Earth's atmosphere, thus invalidating the detection of a global warming signal.

---

[17] https://www.theguardian.com/environment/climate-consensus-97-per-cent/2017/feb/16/scientists-study-ocean-absorption-of-human-carbon-pollution.

[18] https://www.washingtonpost.com/climate-environment/2020/09/14/glaciers-breaking-antarctica-pine-island-thwaites/.

[19] https://www.reuters.com/business/environment/atlantic-ocean-currents-weaken-signalling-big-weather-changes-study-2021-08-05/.

[20] https://www.pnas.org/content/106/35/14778.

[21] https://www.nature.com/articles/s41558-018-0355-y.

[22] https://esgf.llnl.gov/.

[23] https://www.nature.com/articles/s41561-019-0400-0.

[24] https://www.ipcc.ch/report/ar2/wg1/.

[25] https://agupubs.onlinelibrary.wiley.com/doi/10.1029/97EO00351.

The analysis of the satellite data underpinning such "no warming" claims was simply wrong.[26] A lot of hard work had to be done to understand flaws in the satellite data.[27] That hard work advanced the science,[28] ultimately leading to better satellite observations — observations which now show robust warming of the lower atmosphere.[29]

Another popular criticism of the IPCC's 1995 "discernible human influence" conclusion was that it rested primarily on studies of surface temperature changes. Critics argued that if there really was a human-caused climate signal lurking in observations, it should be manifest in many different aspects of climate change — not just in surface thermometer measurements.

This was valid criticism. Climate scientists did not ignore it. You don't get far in any scientific endeavor if you ignore valid criticism.

The response of the scientific community was to examine dozens of different independently monitored aspects of climate change, and to eventually show that human-caused climate fingerprints were all over the climate system[30] — not just in surface temperature. As AR6 reported:

"It is unequivocal that human influence has warmed the global climate system since pre-industrial times. Combining the evidence from across the climate system increases the level of confidence in the attribution of observed climate change to human influence and reduces the uncertainties associated with assessments based on single variables. Large-scale indicators of climate change in the atmosphere, ocean, cryosphere and at the land surface show clear responses to human influence consistent with those expected based on model simulations and physical understanding."

That's how we got from AR1 to AR6. With better data, better models and better understanding of key physical processes. With improved paleoclimate reconstructions. With detailed scientific responses to valid criticism. With individuals willing to devote years of their lives to IPCC assessment reports — a scientific "coalition of the willing" that brought together experts from physics, chemistry, biology, oceanography, meteorology, glaciology, computer science, statistics, social sciences, risk management and many other disciplines.

There will be critics of AR6, just as there were critics of all previous IPCC assessments. The critics will likely opine on Fox News and in the Wall Street Journal[31] about a lack of "maturity"[32] of climate science. The uncertainties are too large, they will argue — the costs of action are too great.

---

[26] https://science.sciencemag.org/content/334/6060/1232.

[27] https://www.nature.com/articles/29267.

[28] https://science.sciencemag.org/content/309/5740/1548.

[29] https://journals.ametsoc.org/view/journals/clim/33/19/jcliD190998.xml.

[30] https://www.ipcc.ch/report/ar6/wg1/#FullReport.

[31] https://www.wsj.com/articles/how-a-physicist-became-a-climate-truth-teller-11618597216.

[32] https://www.wsj.com/articles/climate-science-is-not-settled-1411143565.

Those who deny the reality and seriousness of climate change are wrong. Dangerously wrong. Their incorrect views on the science make it more difficult to achieve real action on climate change. Inaction harms all of us.

The AR6 report diminished the space in which climate change denialism can thrive. Denialism is a tough sell when cautious scientists use words like "unequivocal" in international reports. It's a tough sell in 2021, when climate change manifests in your own backyard, and your backyard[33] is in Siberia, the Pacific Northwest, British Columbia, Greece, Turkey, Algeria, Germany and China. It's tough to sell "Don't worry about climate change" when smoke from uncontrolled forest fires reaches the North Pole.[34]

If you do one thing this weekend, read the Summary for Policymakers[35] of the AR6 report. Understand how our planet's climate is changing. Then act and vote with purpose. Hold political leaders accountable if they fail to address climate change. As the AR6 report clearly states, Earth's climate future is in our hands.

Afterword

Decades of inquiry into every aspect of climate science are summarized in the IPCC assessments. This scientific enterprise is ongoing. There is much to learn about how the planet's climate will respond to continued human greenhouse gas emissions, and how the climate will recover after (hoped) success in eliminating these emissions. Examples of ongoing challenges include reliable projection of climate change effects at regional levels (to inform investments in resilience and adaptation) and better understanding of potential tipping points in the climate system.

In the attribution area, scientists—and legal experts—are increasingly focusing on trying to quantify human contributions to the changing properties of extreme events. Machine learning and artificial intelligence techniques are being applied to extract meaning from enormous volumes of climate model output and observational data. There is a big research thrust in the area of so-called emergent constraints—the search for clever ways of using present-day and historical climate information to reduce uncertainties in projected twenty-first century climate changes.

There is no shortage of interesting and important problems. The next generation(s) of climate scientists still have plenty of work left to do.

---

[33] https://news.yahoo.com/wildfires-rage-across-the-world-as-un-releases-damning-climate-change-report-171925952.html.

[34] https://theweek.com/world/1003585/wildfire-smoke-reaches-the-north-pole-for-1st-time-in-recorded-history.

[35] https://www.ipcc.ch/report/ar6/wg1/downloads/report/IPCC_AR6_WGI_SPM.pdf.

Essay 29. Fighting Climate Change in a Fragmented World

Prologue

Over our careers, we all have participated in programs of climate research and analysis in collaboration with colleagues from around the world. As a result, we are acutely aware of the fact that no country holds a dominant share of the talent needed to understand our complex planet, develop new technological options for emissions reduction, and formulate an effective policy response. Fortunately, the needed cooperation has proceeded smoothly for decades. Examples include university exchanges, funding across borders by government science agencies, and the normal functions of international scientific societies.

In recent years, however, there has been growing tension between the U.S. and China on issues of trade and intellectual property, casting a pall of caution over scientific exchanges. And Russia's attack on Ukraine is leading to a fundamental break in cross-nation collaboration. We wanted to call attention to the damage that such isolation of national programs would impose on the global efforts to tackle climate change. We also tried to point to ways in which a focus on our common interest might be sustained even in the midst of increasing global fragmentation.

This essay first appeared as https://thehill.com/opinion/energy-environment/3479916-fighting-climate-change-in-a-fragmented-world/.

G. Yohe et al., *Responding to the Climate Threat*, https://doi.org/10.1007/978-3-030-96372-9_29

Fighting climate change in a fragmented world

Henry Jacoby, Ben Santer, Gary Yohe and Richard Richels

May 7, 2022

Currently, seven astronauts are living on the International Space Station: three Americans, one European and three Russians. They may not agree on much outside the day-to-day management of their vessel, but they must cooperate on that task if they all are to survive. The same is true for us on spaceship Earth.

Unfortunately, geopolitical forces threaten to dismantle the world order established after World War II. This disruption of relations among the great powers makes it even harder to sustain the international collaboration needed to tackle climate change. Its chilling effect extends not only to cooperative efforts to meet global emissions goals, but also to the research and policy studies that are needed to guide global action.

The growing stress between the U.S. and China is one obvious problem. In advance of the 2015 Paris climate negotiations, President Obama[1] and Chairman Xi met to declare their joint commitment[2] to enhanced climate action. It is hard to imagine a similar meeting taking place today. One U.S. response to the perceived China threat has been enhanced supervision, and in some areas the restriction of academic contacts. The concern lies with the protection of intellectual property in particular technologies, like chips, artificial intelligence and biotech, but the reporting requirements are broad. They cast a pall of suspicion[3] over all joint work, including on climate change. Cooperation has become more difficult[4] for both U.S. and Chinese scholars.

And now there is Russia's unprovoked attack on Ukraine. In response, other nations have taken measures to cut Russia off from the rest of the world economy,[5] remove it from international institutions like the UN Human Rights Council, and isolate Russian citizens, both within Russia and abroad, who support Putin's "special military operation." This now includes not only targeted financial sanctions, but also exclusion from sports, the arts, intellectual exchange and other normal aspects of global society. Considering the extreme brutality[6] of the Russian army's pursuit of the war, it is easy to understand the impulse to punish and perhaps restrain this aggression through every available channel short of expanding the war.

---

[1] https://thehill.com/people/barack-obama/.

[2] https://obamawhitehouse.archives.gov/the-press-office/2015/09/25/us-china-joint-presidential-statement-climate-change.

[3] https://www.nature.com/articles/d41586-019-01270-y.

[4] https://www.nature.com/articles/d41586-020-02015-y.

[5] https://www.cnbc.com/2022/03/11/how-the-us-and-allies-cut-off-russia-from-the-global-economy.html.

[6] https://www.cnn.com/2022/04/04/europe/russia-military-culture-brutality-intl/index.html.

Unfortunately, there is a downside to shutting off all contact with Russian society. Justified as this urge may be, there is a risk of losing sight of our common interest in the midst of conflict. Fruitful collaboration on climate issues involving Russian and U.S scientists and other researchers has essentially ended. Canceling joint projects and suspending communications will have dire implications for the needed climate research[7] and the continuity of supporting data sets.

Nobody knows when or how the Ukraine war will end, or what international trade and security system will evolve from the supply chain disruptions of the COVID epidemic and intensifying U.S.-China rivalry. Some fear the globalized world that has emerged over the last half century will evolve into two or three semi-autonomous trading blocs[8] characterized by mutual hostility and distrust. Such a profound decoupling will make collaboration on meeting the global climate challenge much more difficult.

So what's to be done? Clearly, top priorities include finding a just end to the war and achieving a peaceful resolution of other great power tensions. But there is another task along the way, to preserve international contacts and lines of scientific communication. Severing of these ties will make it more difficult to restore cooperation in what we hope will be a more peaceful and collaborative future.

Thus government agencies, even as they act to protect U.S. interests, need to try to maintain conditions favorable for international climate research efforts. Also, non-governmental organizations, like the National Academies and scientific societies, should oppose calls[9] to sever all contacts with sister agencies. Isolation is not feasible for those on the International Space Station, nor is it feasible to address the global climate threat without Russian and Chinese cooperation.

Action by individual researchers is important as well. Those of us who have been involved in the climate issue should also try to preserve our own personal relations and lines of international scientific communication. Contacts with colleagues in China and Russia are not only critical for reinvigorating collaboration in less adversarial times. They also provide support and encouragement for those brave enough to speak out.

There are dramatic examples of such public acts of courage. Hundreds of Russian scientists signed a letter[10] declaring their unequivocal opposition to the invasion of Ukraine. And the leader of the Russian delegation to a February meeting of the Intergovernmental Panel on Climate Change (IPCC) offered an

---

[7] https://www.reuters.com/lifestyle/science/ukraine-conflict-hurts-russian-science-west-pulls-funding-2022-04-10/.

[8] https://www.bloomberg.com/opinion/articles/2022-03-24/ukraine-war-has-russia-s-putin-xi-jinping-exposing-capitalism-s-great-illusion?sref=B3uFyqJT.

[9] https://www.nature.com/articles/d41586-022-00601-w.

[10] https://www.chemistryworld.com/news/more-than-600-russian-scientists-sign-open-letter-against-war-with-ukraine/4015292.article.

apology[11] for the war. Given the harsh penalties for criticizing the Putin[12] regime, such brave individuals deserve our support — they must not be isolated from the global community of their colleagues.

Indeed, now may be the time to revisit the spirit of the Pugwash Conferences.[13] In the depths of the Cold War, meetings by a group of scientists and policy-makers provided an essential line of communication between the Soviet Union, Western Europe and the U.S. The participants' motivation was to avoid destruction in the then-prevalent nuclear strategy of "mutually assured destruction" (MAD). Efforts to sustain engagement with our Chinese and Russian scientific peers is no less urgent in response to the rising threat of MAD of the global environment.

Afterword

There are encouraging recent examples of common interest among nations overcoming their divisions, even in times of tension. In November 2021, the U.S. and China signed a declaration renewing their joint support for cooperation on climate change response measures. Common concerns even overcame divisions over the Ukraine war. In July 2022, Roscomos and NASA signed a new agreement to give Russian astronauts seats on American rockets and Americans a ride to orbit on Soyuz rockets, at least until the end of the current contract period in 2024. And in response to the very real potential of widespread famine resulting from the Russian blockade of Ukrainian grain exports, Russia and Ukraine came to an agreement in July 2022 to open the Black Sea ports.

It is now common for national leaders to declare that climate change is a "crisis," or an "emergency." We can only hope that this high level of concern leads to protecting international cooperative efforts in research and analysis. These efforts serve our common interest in fighting climate change.

---

[11] https://www.washingtonpost.com/climate-environment/2022/02/27/ipcc-russian-apologizes-ukraine-climate/.

[12] https://thehill.com/people/vladimir-putin/.

[13] https://www.nobelprize.org/prizes/peace/1995/pugwash/history/.

Essay 30. Energy Transformation Can Strengthen Democracy and Help Fight Climate Change

Prologue

Russia's attack on Ukraine, with both predictable and unimaginable consequences, reconfirmed a lesson about these kinds of events that most of us learned long ago: everything is connected to everything else. For example, it's no surprise that the war in Ukraine would impact national climate policies. The race to find new sources of fossil fuels in order to reduce reliance on Russian oil and gas may be a significant setback to progress in fighting climate change. For some countries, war in Ukraine means that emissions pledges under the Paris Agreement could be nearly impossible to meet. The net result is to further delay effective emission reduction, thus wasting time that the planet does not have.

This essay first appeared as https://yaleclimateconnections.org/2022/04/energy-transformation-can-strengthen-democracy-and-help-fight-climate-change/.

A more optimistic possible outcome of the war is an acceleration in the pace of market penetration of existing and future sources of non-fossil energy into global markets. This acceleration, prompted by the urgent need to divest from Russian fossil fuels, preserves the chance of achieving net-zero emissions targets by the middle of this century. It could also reduce the power of fossil-fuel-producing autocracies in the ongoing global contest between democratic and authoritarian regimes. World leaders, particularly in Europe, quickly understood that they faced energy use choices of enormous significance. We wanted to highlight the wider implications of these choices.

Energy transition can strengthen democracy and help fight climate change

Gary Yohe

April 1, 2022

It is impossible to forecast how the war in Ukraine is going to end: current events are fast-moving. Given the inhumanity of it all, it is important to consider the resulting uncertainty and implications for the entire world.

Uncertainty about the global ramifications of the war clearly has driven world prices of liquified natural gas (LNG) dramatically higher[1] over the past several months. These price increases have not hurt Russia: In fact, they have helped to finance its war effort. Rapidly climbing LNG and oil prices, however, have hurt much of the rest of the world, as supplies of LNG have been gobbled up swiftly by the highest bidders with the largest appetites. Those most hurt by all this live in other developed and developing nations all around the world. And even in many European countries and the United States, those with limited means already are suffering.

So what can be done? Any first-year student of economics knows that increasing supplies from all non-Russian sources of energy could work over time, especially in concert with efforts to reduce demand. These are good ideas, of course, but the devil is in the details. There are at least two distinct options:

Option 1: Invest in opening untapped supplies of petroleum and natural gas, drill for more of both, operate existing distribution infrastructure at its fullest capacity, and build more as quickly as possible; or

---

[1] https://www.naturalgasintel.com/european-commission-rolls-out-plan-to-gain-independence-from-russian-fossil-fuels-by-2030/.

Option 2: Two complementary parts, here: (a) invest in expanding diverse and decentralized non-fossil energy systems; and (b) invest in R&D on new technologies that can smooth the demand-side transition to using electricity, technologies such as electric vehicles.

The European Union recognizes that the choice is not binary. The EU's announced plan is designed to reduce dependence on Russian LNG as quickly as possible by expanding access to reserves from the United States, a component of option 1. It seeks to do so[2] while making simultaneous longer-term investments in "frontloading renewable energy and improving energy efficiency" (the very spirit of the dual supply and demand approach of Option 2).

Poland and Belgium already are expanding their LNG terminals, and Greece and Germany have each recently approved construction of three new terminals. Germany has committed to independence from Russian LNG by the middle of 2024. The U.S. has agreed to supply an additional 15 billion metric tons of LNG this year, and the EU will work to promote substitution to LNG to the tune of 50 billion metric tons per year – an effort that will require increased supplies from many places.

But what about the longer term? Details matter there, too. Should the developed world expand the status quo as described in parts of option 1, or should it accelerate its movement toward the environment-friendly structure of option 2? Future investment should favor the latter, and not simply because it would promote a less hazardous climate future. Given the events of the past several decades, it is important to note that doing so would strengthen democracy's place as a fundamental principle of modern government.

Mr. Putin has successfully invaded sovereign nations[3] whenever his hope of resurrecting the old Russian Empire has been threatened by independence movements within former Soviet satellite states. This time, however, he has encountered a country and population not easily subdued. Ukrainians are fiercely and effectively using weapons and training from the West to defend their way of life. Ukrainians have reminded the planet's population that democracy is worth fighting for – to the last breath, if necessary.

---

[2] https://www.csce.gov/sites/helsinkicommission.house.gov/files/Report%20-%20Russian%20Military%20Aggression%20-%20FINAL_0.pdf.

[3] https://www.csce.gov/sites/helsinkicommission.house.gov/files/Report%20-%20Russian%20Military%20Aggression%20-%20FINAL_0.pdf.

Putin's war has pushed world energy markets to inflection points. It has created a perhaps once in a generation opportunity to reorganize and transform global markets toward renewables and thereby reduce the world's dependence on fossil energy from countries with leadership antithetical to democracy (not just Russia). Investing aggressively in energy option 2 would reduce the political power of major fossil fuel exporting nations with authoritarian leaders. Why? Because rapid transition to Option 2 undermines the ability of autocrats to maintain their extraordinary market clench over supplies of scarce and essential commodities. Such a transition would undermine their access to money from the rest of the world – money they use to fund inhumane oppression at home and unlawful and immoral extracurricular aggression abroad.

Shrinking such gains derived from formidable market power would strengthen the hand of democracy – not by making democracy work better (it will always be messy), but by diminishing the use of fossil fuel energy to bankroll wars and hold energy-needy countries hostage. Constraining dictators' and autocrats' power over energy issues can help both to forward democratic principles and to help propel progress toward a cleaner and healthier global environment.

Afterword

Given the long lag times in infrastructure development, it will be very difficult for EU countries and alternative energy suppliers to implement energy choices that will meaningfully reduce dependence on Russia. In a normal year, around 40% of the EU's gas consumption comes from Russia. This huge supply system cannot be expected to change on a dime. In the short run, the larger response is on the demand side. Examples include informational programs (instructing people on how to conserve) while at the same time encouraging "incitement and supporting actions" like rebates on the purchase of energy-efficient appliances.

Over the slightly longer term, the EU Commission put forward options like an "EU Strategy for Solar," designed to double the 2020 supply of solar energy by 2025 with 320 GW of new solar photovoltaic installations, to be supplemented by almost 600 GW of additional supply by 2030. The EU views this action as building "frontloaded capacities" for a different energy future, even though it would only replace 2–3% of current natural gas demand by 2027.

These numbers are small. They show that strong fossil fuel dependence in 2022 will likely continue in the immediate future, even with the war-driven imperative to rapidly transition Europe away from Russian oil and gas.

Essay 31. The Choice Is Clear: Fair Climate Policy or No Climate Policy

Prologue

No matter how you look at it, efforts to confront the climate threat are bedeviled by issues of fairness. These concerns dominate international climate negotiations. In the Climate Convention's 2005 Kyoto Protocol, nations were divided into two groups, developed and underdeveloped, with only the richer group having an obligation to reduce greenhouse emissions. After decades of frustration with the Kyoto system, the Paris Agreement erased this formal grouping, with all nations agreeing to pledge emissions reductions in a common format. Despite this common format, negotiations within the Agreement are still conducted by groups with allied interests. Examples include the G77 (all developing countries), the Less Developed Countries (a sub-group of the G77), and the Umbrella Group (a subset of developed countries).

At the national level, the thorny distributional issues in climate policy development are all too obvious—for example, in conflicts over measures that would impose cost on firms and regions engaged in fossil energy production and use. This is an area where the Biden-Harris Administration has put special emphasis on equity in policymaking. The success of the global climate effort is ultimately dependent on achieving greater equity between nations, both in policymaking and in burden sharing. We felt that this dependence was not widely appreciated. We wanted to explain why it is important for the developed countries, and most importantly the United States, to provide aid to developing nations in their efforts to control emissions and adapt to the serious climate changes they are already experiencing.

This paper first appeared as https://yaleclimateconnections.org/2021/03/commentary-the-choice-is-clear-fair-climate-policy-or-no-climate-policy/.

The choice is clear: fair climate policy or no climate policy

Richard Richels, Henry Jacoby, Benjamin Santer and Gary Yohe

March 22, 2021

President Biden has expressed a commitment to making equity[1] a guiding principle in domestic policy formation. When applied to climate policy, it may help eliminate major obstacles to getting our own house in order. But equity concerns may be a greater barrier when it comes to international negotiations: Poorer countries are demanding that it be an overarching consideration in evaluating policies not only within but across national boundaries.

The image[2] below gets to the nub of the problem. Equality should not be confused with equity. Equality means that we all have access to the same tools and opportunities (ladders of equal height). This approach would be fine if all parties were to have the same access to the same underlying foundations: healthcare, education, jobs, shelter, and the other trappings of wellbeing. The higher ladder corrects for inequality. It gives the disadvantaged a step up, in effect leveling the playing field.

Domestically, one need look no further than the plight of coal miners[3] or autoworkers[4] for examples of what the President has in mind. Moving away from coal and oil will eliminate traditional jobs in these and related industries. The communities where workers live will also be adversely affected. Does America not owe some debt to those who helped power the post-World War II economic boom?

Then there are the negative impacts from rising energy prices on disposable income. These impacts will be particularly hard-felt by those who are but one paycheck away from poverty. Through no fault of their own, they will bear a disproportionate burden from our past political recalcitrance.

The issue is how to cushion the blow. Policymakers have a number of tools at their disposal, ranging from market-based instruments to so-called command-and-control approaches. These can be used not only to discourage fossil fuel use, but also to address ensuing issues of equity. For example, with market-based instruments (e.g., carbon taxes, and cap and trade), the resulting revenue can be recycled back into the economy in a manner that compensates those most disadvantaged.

---

[1] https://www.whitehouse.gov/briefing-room/presidential-actions/2021/01/20/executive-order-advancing-racial-equity-and-support-for-underserved-communities-through-the-federal-government/.

[2] https://www.slideshare.net/GabrielGaldamez/tony-ruths-equity-series-2019-247618162.

[3] https://www.edf.org/blog/2018/05/08/why-these-companies-turn-coal-country-skilled-workers#:~:text=From%20coal%20fields%20to%20solar%20rooftops%20and%20coding.

[4] https://www.washingtonpost.com/climate-environment/2021/02/06/auto-industry-peers-into-an-electric-future-sees-bumps-ahead/.

Facing up to past recalcitrance

A command-and-control approach is more prescriptive. In this case, redress would lie in the hands of government agencies to fashion programs to retrain affected workers, compensate impacted communities, and help those most harmed by higher prices at the gas pump and in their monthly utility bills.

Most likely a combination of the two approaches will be needed. Addressing issues of equity will not be easy, however, and there are many political "third-rails." Matching the supply and demand for jobs will entail careful coordination between the private and public sectors. The effects on surrounding communities will be complex. And even a small redistribution of income will likely be met with fierce resistance. Bipartisan leadership, sorely missing in recent years, will be required. But the alternative is unacceptable – inaction on the climate change problem, a "lose-lose" situation for all.

Global needs raise complex challenges

The meaning of fairness is further complicated at the international level.[5] The first wave of the industrial revolution was powered by cheap and abundantly available fossil fuels, with planetary warming an unintended by-product. Now those aspiring to a standard of living similar to their wealthier neighbors are being asked to abandon fossil fuels in favor of cleaner but perhaps more expensive alternatives.

Again, the issue is how to cushion the blow. The United Nations Framework Convention on Climate Change has called upon[6] wealthier countries to help their poorer neighbors in the transition to a carbon-free future. This action would require transfers of capital (financial and technological) to other countries. Unless agreement can be found on what constitutes equitable burden-sharing, international negotiations may grind to a halt.

For those who say we will just wait them out, bear in mind that climate change may not be as high on some countries' lists of priorities. Particularly, the poorest who understandably are likely to be concerned with more pressing worries such as immediate survival.

Then there is also the issue of our access to adaptation possibilities.

Moreover, there is sufficient warming[7] baked into the climate system to cause considerable harm. In adapting to these inevitable changes, how should the pain from past procrastination be distributed? Many countries lack access to adaptive capacity that the wealthier can muster. For example, they cannot build multi-billion-dollar sea walls to safeguard against rising oceans in order to protect life, property, and valuable ecosystems. Nor do they have the resources[8] to ensure widespread deployment of vaccines and needed therapeutics to guard against climate-sensitive diseases such as malaria, yellow fever, and dengue and Zika viruses.

Furthermore, many people live in countries where the political leadership is struggling to maintain civil order. Only limited resources are available to meet the most basic needs of the population of a failing state. Climate change can significantly exacerbate such existing political instability, potentially leading to the development of regional conflicts and hundreds of millions of environmental refugees.[9] In these circumstances, controlling greenhouse emissions falls even lower on national priorities.

---

[5] https://www.google.com/search?q=Ringius%2C+L.%3B+Torvanger%2C+A.%3B+Underdal%2C+A.+(2002)&oq=Ringius%2C+L.%3B+Torvanger%2C+A.%3B+Underdal%2C+A.+(2002)&aqs=chrome..69i57.2412j0j9&sourceid=chrome&ie=UTF-8.

[6] https://unfccc.int/resource/climateaction2020/.

[7] https://www.nap.edu/resource/12877/Stabilization-Targets-Final.pdf.

[8] https://www.ipcc.ch/report/sixth-assessment-report-working-group-ii/.

[9] https://www.brookings.edu/wp-content/uploads/2019/07/Brookings_Blum_2019_climate.pdf.

So, whether it is reducing global warming or ameliorating the harm inflicted by past intransigence, we face daunting challenges. Equity must be a primary concern not only in domestic, but also in international policy formation. The expression "helping thy neighbor" is usually reserved for our fellow citizens, but if developed countries fail to help poorer countries, the global effort to control warming will falter. Close attention to this issue by the Administration and the Congress is essential. Equitable climate policies are not only the right thing to do – they are also in our own national self-interest.

Afterword

Thus far, assistance provided by the wealthier countries does not live up to that required to enable an effective global response to climate change, or even to meet pledges they have made over the years. For example, in 2009, wealthier nations agreed as a group to provide developing countries with $100 billion in climate-related aid per year by 2020. It could be contributed through a number of different channels: bilateral agreements, international development banks, or in payments to funds organized under the Climate Convention. The $100 billion is a small number in relation to the need, but to date the totals fall below even this modest goal.

Support has been similarly disappointing for direct national contributions to funds organized under the Climate Convention, like the Green Climate Fund which finances emissions mitigation and climate resiliency in developing countries. The U.S. in particular has fallen far behind on its obligation. The Trump Administration zeroed out support for the Green Climate Fund. Even under the Biden-Harris Administration, which pledged a renewal of U.S. contributions, Congress has been reluctant to approve the funds. It is a hard sell—that the U.S. must not only reduce its own emissions but also raise the money to help others reduce their emissions. But it is a contribution that the American public needs to accept if the U.S. is not to be a persistent drag on the global effort to avoid dangerous interference with Earth's climate.

Essay 32. The 1.5 Degrees Goal: Beware of Unintended Consequences

Prologue

In writing the essays that make up this book, part of our job was to inform the broader public about the climate change threat and the urgency of action to cut global greenhouse emissions. We are sympathetic to the efforts of others to encourage people to pay greater attention to climate change, and to send a clear signal that this should be a high-priority issue for the government. But we became concerned by some of the language being used by public officials and in the press—language that characterizes the climate threat in ways inconsistent with available science. Credibility is just as important for those communicating about climate change as it is for the scientists studying it. Hyping the threat beyond the current scientific understanding risks loss of trust and unnecessary despair.

This essay first appeared as https://yaleclimateconnections.org/2022/01/the-1-5-degrees-goal-bew are-of-unintended-consequences/.

The 1.5 degrees goal: Beware of unintended consequences

Richard Richels, Henry Jacoby, Benjamin Santer, and Gary Yohe

January 5, 2022

"Keep 1.5 alive" emerged as the haunting refrain of the recent United Nations climate conference in Glasgow.[1] Although a well-intentioned rallying cry, it raises important questions about how the chant is to be interpreted. Unfortunately, 1.5°C is often presented as an immutable crisis point, rooted in established scientific consensus. It's implied that beyond this point, climate-induced damages increase dramatically.

Given the prospect that the 1.5°C target may not be met, proponents might come to rue their choice of mantra. Not only will it likely cause unnecessary despair, but oversimplification of the underlying science provides those resolutely opposed to acting on climate change with opportunities for further mischief.

This is not to argue that danger points don't exist. There is ample evidence in paleoclimate records of sudden and dramatic shifts in Earth's climate system, though the conditions that triggered them are not fully understood and are not directly comparable to present-day conditions.

Another reason for concern comes from computer models of the climate system. Models also alert us to the possibility[2] that continued warming may cause rapid climate changes that cannot be easily reversed for centuries or longer. Examples of such changes include abrupt sea-level rise and slowing or even shutdown of a key part of the ocean's system for circulating heat. While few studies suggest that these changes are imminent, models and "deep time" climate records both tell us that somewhere out there, beyond the 1.1°C of warming experienced to date, dangerous tipping points exist. We don't know exactly where they are, just that further warming makes it more likely we will cross them.

This understanding often contrasts sharply with the statements made by public officials and the press. For example, United Nations Secretary-General Guterres implicitly attributes[3] the politically-set 1.5°C temperature target to the science. He refers to the 2021 release of the Physical Science section[4] of the latest IPCC Assessment Report as a "code red for humanity" and notes that "The internationally agreed threshold of 1.5°C is perilously close." But nowhere does the IPCC special report[5] on the impacts of 1.5°C warming suggest that a critical, well-established threshold exists at that precise level of warming: It simply

---

[1] https://ukcop26.org/cop26-keeps-1-5c-alive-and-finalises-paris-agreement/.

[2] https://media.nature.com/original/magazine-assets/d41586-019-03595-0/d41586-019-03595-0.pdf.

[3] https://www.un.org/press/en/2021/sgsm20847.doc.htm.

[4] https://www.ipcc.ch/report/ar6/wg1/.

[5] https://www.ipcc.ch/sr15/chapter/spm/.

concludes that limiting warming to 1.5°C, if it could be achieved, would be better than limiting warming to 2°C.

Likewise, climate stories in the news media[6] commonly refer to a 1.5°C critical threshold beyond which the planet will experience increasingly deadly floods, wildfires, and enormous storms – and then add "scientists say." Crediting the climate science with establishing a specific damage threshold at 1.5°C of warming, which scientists did not do, risks erosion of public trust in their work. It also provides ammunition for those who would criticize the scientific enterprise[7] for hyping the issue, or for using scare tactics that go beyond the available evidence.

Limiting global warming to 1.5°C is a daunting task. The climate system has not fully adjusted to the levels of greenhouse gases currently in the atmosphere, so some additional warming is already baked in,[8] even if all emissions were to cease overnight.

There are also inherent impediments to how quickly existing energy-producing systems and capital stock (power plants, transport, and buildings) can be replaced or renovated to create nonpolluting alternatives. Together, these lags – in both the physical and human systems – suggest that 1.5°C may well already be in the rearview mirror.

Strong language, metaphors understandable ... but carry risks

These are dramatic times for Earth's climate. It is not surprising that those trying to spur public effort to counter the real and serious threat of global climate change use dramatic language and metaphors, try to anchor political targets in scientific proof, and stress frightening consequences of inaction. But motivating action on climate change is a delicate undertaking.[9] While the scale of coming climate damage clearly calls for urgent action to transform the energy economy, there are risks in painting too bleak a picture of the challenge, and in presenting 1.5°C as the well-established climate equivalent of an end-of-the-world prediction.

One must be aware of the possibility of unintended and unwanted consequences: Paralysis and despair may arise if millions believe that exceeding the 1.5°C target inevitably signals climate Armageddon, beyond which all is lost. Such despair would imperil the continued energy and attention needed to sustain the global effort to cut greenhouse emissions.

While useful as a spur to action, "Keeping 1.5 Alive" must not be allowed to obscure the fact that it's worth fighting to prevent every 0.1°C of additional warming – up to and (importantly!) beyond 1.5°C. And that every 0.1°C of warming avoided is cause for celebration and hope.

---

[6] https://www.nytimes.com/2021/05/18/climate/climate-change-emissions-IEA.html.

[7] https://nypost.com/2021/04/24/obama-admin-scientist-says-climate-emergency-is-based-on-fallacy/.

[8] https://doi.org/10.1073/pnas.0812721106.

[9] https://undp.medium.com/talking-about-climate-change-walking-the-line-between-hope-and-despair-6605cf757ba5.

Afterword

As this book neared completion in the summer of 2022, it appeared ever more likely that global temperature will cross the politically determined "red line" of 1.5°C within the next several decades. The notion of a scientifically well-defined threshold or "cliff" of climate damage at that specific level of warming continues to be advanced by advocates of urgent action. The existence of such a cliff also is frequently reported in press stories. Sadly, current science cannot precisely pin down the levels of warming at which we pass certain thresholds.

That said, it's also inarguable that even at the roughly 1.1°C temperature increase we've experienced to date, serious, life-threatening impacts have already occurred. These impacts include the "NB4" extreme heat events discussed in a previous essay (Essay 27), along with devastating droughts and horrific wildfires. It is increasingly important to convey the understanding that motivating action to reduce greenhouse gas emissions does not require false certainty about the degree of warming that triggers serious and impactful changes in Earth's climate. And it's important to let people know that anything we can do to cut emissions will help to prevent the worst outcomes.

Essay 33. A Durable U.S. Climate Strategy … or a House of Cards?

Prologue

So how do we clean up the mess we have created? The preponderance of evidence gathered over decades of research and observation suggests that nothing less than a transformation of the energy system will be required, likely accompanied by major shifts in lifestyle. There is not widespread public understanding of the fact that a change of this magnitude will require continuity of government policy over decades, even generations. Climate goals are not achievable if, every time we gain some momentum on the issue, momentum is overwhelmed by the political exigencies of the moment.

The word that seemed to us to be missing in policy debate is "durable". We saw the need for a clear statement of what will be required to get off a broken treadmill, where one step forward is too easily followed by one step back, or worse.

This essay first appeared as https://yaleclimateconnections.org/2022/06/a-durable-u-s-climate-strategy-or-a-house-of-cards/.

G. Yohe et al., *Responding to the Climate Threat*, https://doi.org/10.1007/978-3-030-96372-9_33

A durable U.S. climate strategy ... or a house of cards?

Richard Richels, Benjamin Santer, Henry Jacoby, and Gary Yohe

June 6, 2022

Just when it seemed that real progress might be made on climate change, war has pushed climate concerns to the back burner.[1] This shift in focus is understandable. War is affecting millions in Ukraine and poses a growing threat to global security. But it's dangerous to leave the climate change problem untended, simmering away on the back burner. Like war, human-caused warming also poses an escalating threat to human lives, livelihoods, well-being, and the stability of democratic systems of governance. Things left out of sight and untended on a hot stove can combust.

Our global society does not have the luxury of being able to focus on only one threat at a time. Threats are intertwined, synergistic. Today's petro-dictators, using oil money to finance war, are intent on enhancing rather than diminishing reliance on fossil fuels. Fossil carbon is their life support system. Its continued use also adds to emissions of heat-trapping greenhouse gases, imperiling our planetary life support system.

The bottom line is that we must pay attention to the many threats to human wellbeing boiling over on the war burner, the global pandemic burner, and the climate change burner. And now, at least for the U.S. in the aftermath of the horrifying string of mass murders in Buffalo and Uvalde (list could go on), we cannot forget the "guns" burner. The price of freedom, human health, and planetary health is constant vigilance on these and many more issues.

Stopping greenhouse gas pollution will require a complete transformation[2] of the way the global community produces and uses energy. We cannot achieve that goal without sustained efforts on many fronts: technological, scientific, socioeconomic and political. In a world beset with multiple threats, the challenge is to carve out a durable climate strategy – one that can withstand the distractions of other critical, but inevitably shorter-term, crises.

Climate change is a crisis that will be with us for centuries.[3] This sad but undeniable reality is one of the critical legacies of the long lifetimes in the atmosphere of the greenhouse gases emitted by fossil fuel burning. It will take centuries to millennia for the climate system to come into balance with atmospheric concentrations of greenhouse gases. Our actions are committing the globe to a future, warmer climate that humankind has not previously experienced.

---

[1] https://www.theguardian.com/environment/2022/may/10/john-kerry-warns-long-ukraine-war-threaten-climate-efforts.

[2] https://www.iea.org/reports/net-zero-by-2050.

[3] https://www.google.com/books/edition/Climate_Stabilization_Targets/0j9VfGJ9hvQC?hl=en&gbpv=1&printsec=frontcover.

So what constitutes a durable, effective strategy for dealing with human-caused climate change, preventing it from being relegated to the back burner? Our concern with this question is focused on the United States. There are multiple reasons for this. The U.S. is where we live, and is responsible for the greatest share of global emissions of greenhouse gases since the mid-1800s. Additionally, America's failure to enact durable climate policy inevitably will encourage similar inaction by other nations.

Despite the current Washington stalemate on climate, it is important to keep in mind the necessary attributes of a durable climate strategy.

Elements of making our climate strategy durable

First, the components of a strategy must be credible. Targets and timetables need to be carefully examined through a feasibility lens. If unduly ambitious, they provide ammunition for policy opponents and risk the loss of public trust. For example, the Biden Administration has called for electric vehicles[4] to represent 50% of auto sales by 2030. This target is unlikely to be achieved given that the U.S. is not installing the necessary charging infrastructure, strengthening supply chains, and instituting needed subsidies to support such a rapid shift in consumer behavior. An enterprise of this magnitude will require periodic reality checks to monitor progress along the way, and the same is true for each element of a U.S. climate strategy. Is it on track? Are course corrections necessary?

Next, once adopted, policy measures need to have some sticking power to withstand inevitably shifting political winds. Durability will require avoiding the 'ping pong' phenomenon in which recent administrations have used executive orders,[5] only to see them reversed by a following administration. The obvious solution is congressional legislation, which is not as easily reversed. Politics is often said to be "the art of the possible": A crucial task in developing a durable climate strategy is to find some way of bringing climate legislation into the realm of the possible.

But politicians must enjoy broad public support for such legislation. Constituents must come to see the connection between the lawmakers' proposals and public well-being. An important aspect of this "connecting the dots" will involve communicating the value of investments in both climate science and mitigation technology. Improved understanding of the workings of the climate system can lead to increased damage avoided. And technology research, development, and deployment offer the prospect of bringing down the costs of carbon-free replacements.

It is not enough to say that benefits of reduced pain and suffering justify the costs. Attention needs to be placed on how individual groups may be affected. This applies especially to the issue of who bears implementation costs. For

---

[4] https://www.whitehouse.gov/briefing-room/statements-releases/2021/12/13/fact-sheet-the-biden-harris-electric-vehicle-charging-action-plan/.

[5] https://www.washingtonpost.com/opinions/playing-presidential-ping-pong-with-executive-orders/2014/01/31/cbb6fe30-89f3-11e3-a5bd-844629433ba3_story.html.

example, suppose an emissions pricing method like tax and dividend were to be used. Such systems can be designed so revenue is returned to those at lower income levels, used to repair the social safety net, retrain those who lose their jobs, or support new manufacturing facilities in depressed communities. A strategy that includes close federal/state cooperation is a great asset here: The states provide rich laboratories for testing human response[6] to alternative incentive and rebate systems.

Another key attribute of lasting climate policy is that decisions made today should not be engraved in stone. They must be constantly reevaluated and tweaked as the scientific community learns more about the climate system, the impacts of climate change, and the effectiveness of different policy solutions. Durability must allow for learning, for recalibrating, for updating, and for flexibility when appropriate.

Finally, durability may not be in everyone's interest, and mastering the art of climate politics demands greater transparency. There is no monolithic "public" to which politicians are beholden – elected representatives are responsive to many different publics. For example, constituents who own fossil fuels may wish for continued procrastination, while others deeply concerned by climate change threats want urgent action. Satisfying one constituency likely means alienating others, so politicians may wish to thread the needle by avoiding too much candor.

But citizens need to know where their elected representatives stand, and we should demand that every elected representative goes on the public record regarding climate strategy. Do they have a plan? Are their plans credible? Do they seek bipartisan solutions? Are they paying attention to the science? If not, let them say so or remain silent. Their silence is an answer – a sign that they do not treat the threat seriously.

In the United States, decades of climate policy have been like a house of cards. It's easy to begin building, but the house is difficult to expand and maintain. It's highly vulnerable to slight shifts in political winds. We need to build a better, stronger house. Our kids will be living in it for a long time to come. We owe it to them and to future generations.

Afterword

It turns out that early June 2022 was an appropriate time to hoist a flag in favor of durable measures to confront a long-term challenge like climate change. In that month, the U.S. saw decisions both consistent with durable policy and at odds with durable policy. As we argue in the essay, legislation conveys durability, and later in the summer of 2022 Congress passed the Inflation Reduction Act—a bill including large-scale federal expenditures on clean energy. That was a win for durability.

---

[6] https://nj.gov/rggi/docs/rggi-strategic-funding-plan.pdf.

On the other hand, one of the more solid components of U.S. environmental policy, importantly including potential climate measures, is the 1970 Clean Air Act. A June 2022 decision by the Supreme Court severely limited application of the Act to greenhouse gas emissions control without further clarification from Congress. This was a loss for durable climate policy.

Part VII The Work Ahead

The Daunting Task

The subtitle of this book is not an exaggeration: climate change *is* society's greatest challenge. We are still in the early stage of a multigenerational struggle to eliminate greenhouse gas emissions and to adapt to the warming to which these emissions have committed us. Dealing with this problem will demand the attention and effort of our children and grandchildren. And yet inertia in the climate system, and in the global economy, means that urgent action is needed now. With each additional day of significant emissions of heat-trapping greenhouse gases, we are committing the planet to warming that will be difficult to reverse on timescales of a few human generations.

Meeting this challenge requires an all-society effort: natural and social scientists, elected representatives and government agencies, international institutions, non-governmental organizations, and university researchers. And most importantly, support from a concerned public. Climate change is but one item in a long list of pressing concerns. Needed political commitments are highly unlikely to get far ahead of the priority that the public places on this issue.

Essential to meeting this daunting task is a public, and policymakers, who are well informed about the climate change threat and its possible solutions. In turn, accurately informing the public and policymakers requires supportive public media outlets that base their output on sound scientific research and analysis. Relevant research and scientific assessment reports have been available for decades. Since the 1970s, climate scientists have written, spoken,

© The Author(s), under exclusive license to Springer Nature
Switzerland AG 2023
G. Yohe et al., *Responding to the Climate Threat*,
https://doi.org/10.1007/978-3-030-96372-9_34

and testified about their findings. Natural and social scientists have published comprehensive studies of the economic, environmental, and social effects of climate change, and of the likely outcomes of a wide range of alternative policy measures.

The essays in this volume contribute to this long effort. We have attempted to gather the complex, interlocking pieces of the climate change puzzle produced by the physical and social sciences, technology, engineering, and economics. Further, our objective has been to assemble these individual pieces into a coherent picture of the causes and impacts of climate change, and to describe this picture of the climatic "shape of things to come" in plain English.

This communication challenge is ongoing. Effective communication must be nimble, quick to exploit "teachable moments," and adaptable to the constraints and opportunities of an ever-changing science and policy landscape. And even in today's time of "never seen before" extreme events (see Essay 27), communicating about climate science still faces the obstacle of well-funded efforts to hamper public understanding of the reality and seriousness of climate change.

In the following, we present a few final thoughts about the road that lies ahead for climate science communication.

Staying the Course

Explaining how science works Enhancing public understanding of how the scientific community works as it seeks to understand our planet is first among the ongoing tasks. Without public understanding of the scientific process, it's difficult to build public trust in the credibility of climate science. Building public confidence has to involve communication at the right level—a deeper level than the coverage available in most daily news feeds, but a level that doesn't get too far into the technical details of the science and policy literature.

Describing the nature of the challenge Hammering home the fact that climate change is the mother of all "tragedy of the commons" problems (as noted in Part I) is another urgent need. Even for a rich, technologically advanced country like the U.S., climate change poses significant risks.

Fossil-energy-driven economic development has made an invaluable contribution to human life and society, with the greatest benefit accruing to wealthy, developed countries. But the resulting emissions of heat-trapping gases to our common atmosphere are committing the globe to warming that damages all. The pain is likely to be greatest in less-developed countries. Halting greenhouse gas emissions requires a transition away from fossil energy, and abandonment of infrastructure been built up over a century and more. It will be a wrenching change that must be accomplished by all nations, rich and poor.

The very nature of a commons problem is that no nation can solve it alone—nor is there is an international police officer to enforce action. Progress therefore relies on voluntary commitments by sovereign nations. These types of pledges are unlikely to be sufficient to meet the goal of limiting warming to under 1.5°C or 2°C unless the burden of reducing emissions is perceived to be equitable. A disinterested and disengaged U.S. has, in the recent past, made an equitable burden-sharing agreement far more difficult to achieve—yet without such an agreement the U.S. will also suffer serious climate change impacts.

The challenge, therefore, is to convince the U.S public and its representatives that it is in the best interests of the U.S. to show real leadership over the long term, or at the very least to take meaningful action in tackling the climate change problem. The richest country and largest contributor to the problem cannot afford to sit on the sidelines.

Emphasizing urgency Beyond the push for climate action, explaining why the need for action is urgent, even in the face of uncertainty, is essential. In our view, this explanation best involves establishing climate change as a challenge of risk management. This is a familiar challenge. We all face decision-making under uncertainty in our daily lives.

True, climate change is a particularly complex puzzle of science, economics, and policy, and all the pieces are never all at hand. But we do not have the luxury of waiting for all the pieces. Though new data and analyses continue to fill in a puzzle piece here and there, wise decisions depend on the best available information *in the moment*. As new information becomes available, we can fine-tune decisions and directions, but it's sheer folly to wait for perfect understanding before acting.

To foster understanding of why delay is costly, continuing effort is required to correct a common under-appreciation of the "cumulative" character of the problem. Emissions today increase the greenhouse gases resident in the atmosphere, committing the globe to more future warming. Long-lived investments in the technology associated with fossil energy supply and use are being made each day, and the increase in fossil infrastructure will affect greenhouse gas emissions for decades. Current decisions have long future tails, creating future damage and pain that cannot be reversed.

Understanding the stakes Finally, on the theme of continuing needs, it's critically important to confront people with solid and credible information about what is at stake (see Essays 16 and 17). Nature is helping, of course. Public perceptions of a serious climate threat are heightened by extreme weather events. But public understanding of this threat must also be based on reliable and accessible estimates of the likely outcomes if the world's nations fail to act effectively. This includes, for example, documenting the loss of valuable ecosystems, disruptions to economies and societies, and the resulting toll on human health and well-being. There is also a need for clarity about the imperative to be well-prepared to face the more hostile climate that is unfolding. Without coherent strategies to bolster climate resilience, climate change will exact an economic and social toll far beyond anything we've experienced thus far.

Broadening the Scope

Preparing for surprises Climate change is a dynamic issue, presenting new challenges and opportunities as scientific understanding evolves. The science is never "done and dusted." There will always be surprises—like the extraordinary heat waves and floods of the 2021 and 2022. While climate scientists have long anticipated changes in such extreme weather in response to human-caused warming and moistening of the atmosphere, the size and frequency of these events has surprised even those who spend their lives studying these issues.

Scientists are now unpacking this surprise and are investigating whether warming causes atmospheric and ocean circulation changes that might favor the development of weather conditions conducive to heat waves. Such circulation changes could partly explain why the severity of recent "real-world" heat waves has outpaced computer model predictions.

More such surprises are guaranteed. We are probably witnessing, in real time, surprises in the feedback loop between warming, the area burned by forest fires, and the injection of carbon dioxide and soot aerosols into the atmosphere. These surprises are an opportunity to educate the public about the dynamic nature of science.

We are only beginning to grapple with a different class of surprise—the potential interactions of global warming with other environmental and social issues. The COVID pandemic provided a wake-up call. Although epidemiologists had warned of the threat of new and dangerous viruses for decades, little attention had been given to how climate change may accelerate their occurrence and amplify their spread (see Essay 10).

Another sobering example of this type of interconnectedness is the relationship between the structure of the energy supply system, climate change, and geopolitical conflict. Recent Russian adventurism would have been more difficult without its large reservoir of hydrocarbons to finance its incursion into Ukraine. Decreased dependence on fossil fuels will not only reduce global warming—it will also lessen the influence of fossil-rich oligarchies (see Essay 30).

Cutting through the fog Encouragingly, a global political response to the climate threat is emerging. The Paris Agreement has led to highly publicized national pledges under its procedures. Countries must formally declare their ambitions to reduce greenhouse gas emissions—by how much, and by when. They have to go on the record, which affords an opportunity to be held accountable. This process is not ideal, but it is gradually ramping up the pressure to implement emissions reduction policies at the national, state, and local levels.

But the prospects for success in meeting temperature targets are unclear. While the Paris Agreement and the U.S. Inflation Reduction Act are grounds for cautious optimism, climate programs have been buffeted by war, geopolitical tensions, and political upheaval. The public will need assistance in interpreting the consequences of these events. And citizens require help to cut through the fog of complex, interacting actions and proposals. There is an urgent need for sober, non-partisan analysis of these proposals—for appraisal of their likely effectiveness in cutting emissions, their likely costs, and (where appropriate) an evaluation of their impacts from the perspective of environmental justice (see Essay 31).

Thinking about adaptation The increase in the number and severity of extreme and damaging weather events is already amplifying discussion of investments in adaptation. Local, state, and federal expenditures—and regulatory measures like mandatory retreat from the coasts—have been proposed in response to the projected changes in climate. We expect to see increasing calls for public expenditure to help cities, firms, and individuals affected by fire, drought, and flooding associated with climate change. As these costs continue to mount, the viability of subsidized insurance and after-the-fact FEMA aid will be up for serious debate, along with the question of who bears the ultimate responsibility for covering the costs of insurance and climate damage. In an ideal world, adaptation decisions would be informed by best estimates of the expected size and rate of projected changes in climate. That is yet another communication opportunity for folks like us.

Maintaining U.S. attention and commitment Lastly, despite efforts of the current Biden-Harris Administration and Congress, international confidence in the U.S. as a solid supporter of the global climate effort remains shaky at best. Having negotiated two climate agreements and then rejected each (permanently from the Kyoto Protocol and temporarily from the Paris Agreement), the durability of the U.S. commitment to solving the climate problem is suspect—particularly considering that one of the two major U.S. political parties has opposed emissions control legislation, sued to block executive actions, and opposed financial aid to developing countries.

The deep U.S. political division on climate policy is due to factors far beyond inadequate understanding of the climate threat and potential solutions. But it is nonetheless urgent to at least try to close any knowledge gap that might contribute to the political divide.

And for Us...

During our careers, all four of us contributed to the task of advancing climate science. We worked on small pieces of the overall puzzle—on improving understanding of the climate system, on risk assessments, on integrated assessments—and we participated in policy analyses. But now, it seems to us that our most valuable contribution may be to participate in public discourse on the reality and seriousness of climate change. To keep talking and writing about the impacts of climate change on things we care about. To help keep discussion of climate change solutions on the front burner of the stove of public opinion.

This book is our contribution so far to that ongoing discussion. It's not an end to our contributions. We hope that it will help you to continue yours.

Printed in the United States
by Baker & Taylor Publisher Services